Dehumanization of Warfare

Wolff Heintschel von Heinegg • Robert Frau •
Tassilo Singer
Editors

Dehumanization of Warfare

Legal Implications of New Weapon Technologies

Springer

Editors
Wolff Heintschel von Heinegg
Faculty of Law
European University Viadrina
Frankfurt (Oder), Germany

Robert Frau
Faculty of Law
European University Viadrina
Frankfurt (Oder), Germany

Tassilo Singer
University of Passau
Passau, Germany

ISBN 978-3-319-67264-9 ISBN 978-3-319-67266-3 (eBook)
DOI 10.1007/978-3-319-67266-3

Library of Congress Control Number: 2017960797

Printed on acid-free paper

This Springer imprint is published by Springer Nature
The registered company is Springer International Publishing AG
The registered company address is: Gewerbestrasse 11, 6330 Cham, Switzerland

Preface

Over the last decades, conventional warfare has evolved into a battlefield of high technology. Drones, weaponized robots, swarms of ships, and the bundling of all ingoing information into a multilayered unified military intelligence picture contribute to warfare, which happens faster than ever before. Computer technology and cyberspace are not only used for command and control of these unmanned systems but for operations of their own, which are undertaken more clandestinely in both worldwide information systems and their interwoven networks. As today's world is becoming increasingly more interconnected and dependent on cyber/information technology, the possible impact of cyber attacks is growing enormously.

The challenges of technology in its relation to mankind and their interrelations, however, are an old story and have been analyzed in depth before. In this area of conflict, law and ethics are of great importance. Every time new technology is getting into conflict with mankind by imposing the risks of damage, suffering, and victims, it is the function and responsibility of law to regulate this technology and limit its future use. The interplay of war and technology has also been discussed at least since the nineteenth century and the industrialization of warfare. Besides the rise of weapons of mass destruction, the discussion got a new momentum due to the exponential growth of abilities of computer technology and its utilization in warfare. The possible interrelations and thereby the possible areas for conflicts between these goods are increasing constantly. Accordingly, the question concerning the need for regulation of new weapon technology is always present. As the consequences and risks for elementary goods of mankind are at their greatest in war, warfare was, is, and will always be the most controversial field in which mankind, technology, and law collide.

That said, the segregation of the human individual out of this triangle of technology, warfare, and mankind is a rather new topic. Warfare solely performed by technology itself is a less researched field. The removal of human beings out of acts of warfare itself is at the peak of the current discussion about warfare in the twenty-first century. A trend of this "new" warfare is to foster more and more autonomy of weapon systems in order to diminish the dependence on endangered, slow, and expensive humans in the battlefield. What unmanned systems and cyber operations have in common is that the human is removed from navigation, control, and decision-making processes. The battlefield of the future, therefore, will contain fewer and fewer human individuals on the ground and at the center of the battle. This development raises questions on multiple levels particularly from a legal and an ethical perspective. For example, the question whether a machine should be able to initiate a process which leads to the death of a human is a highly controversial issue at the very core of the (ethical) discussion. From a legal point of view, the elementary question is whether existing law is still suitable to regulate these upcoming challenges of autonomy.

Our approach to the discussion on autonomous weapon systems and cyber warfare technology was to investigate whether these two separate developments might have something in common that can be utilized for a comprehensive legal analysis. Both methods of warfare involve reduced human control of ongoing actions. Therefore, the generic term "dehumanization" was chosen to express this deficiency of human control. By using the notion "dehumanization," the focus of the legal analysis was determined to examine whether or not the dehumanization of warfare poses legal challenges. In our understanding, this would be the case if the dehumanization was not recognized in the current law and/or existing law was not suitable to regulate all consequences of dehumanization. As a first step, we wanted to bring together different approaches and legal views on the topic by organizing an international conference. However, a purely legal discussion would miss the fairly important technical backgrounds and the ethical and practical aspects of dehumanization, which would not be insufficient to tackle the high complexity of this matter. Hence, we decided to organize an interdisciplinary conference which would involve international experts in the fields of ethics, computer science, unmanned systems, and international law and which would involve decision-makers and members of relevant international organizations like the ICRC. After the award of a grant by the German Peace Foundation in autumn of 2014, the conference was held in February 2015 at the European University Viadrina in Frankfurt (Oder), Germany.

The overall research objective of the conference was to identify possible challenges by unmanned warfare and cyber operations, their commonalities, and their possible consequences for a legal analysis. Additionally the goals from a legal perspective were to determine the legal framework governing the dehumanization and to analyze the identified rules for legal challenges and thereby reach a first result on how to judge dehumanization legally and whether the term itself is suitable for these purposes.

The central focus of the conference was on international law, international humanitarian law, and international criminal law and on ethics, philosophy, and political science. Key issues from the legal perspective were Article 36 of Additional Protocol I, the general principles of humanitarian law, and the notion of direct participation in hostilities. As main topics, the conference focused on autonomous weapon systems, cyber warfare, and unmanned vehicles. These two types of means and methods of warfare were connected by the concept of the human involvement in warfare or the decrease of the same.

The leading questions were the following: Which provisions are relevant for the dehumanization? Which provisions are relevant for unmanned systems and cyber warfare? Are these the same legal rules? Where do dehumanization, unmanned systems, and cyber warfare pose problems to existing law? Can ethical, moral, and political aspects be included in the (understanding of the) law? If so, how? Does the law prescribe that an action has to be performed by a human? Is the involvement of the human (or at least the degree of the involvement of the human) a decisive criterion—from the perspective of law, ethics, or political science? Is a different or modified interpretation of the law required due to any of these issues? Does the existing law govern dehumanization effectively?

The conference was divided into five panels. The first panel served as an introduction to dehumanization from the perspective of international humanitarian law. The framework of international humanitarian law was presented, and more general aspects of dehumanization such as autonomous weapon systems and autonomous targeting were discussed. The second panel focused on interdisciplinary findings concerning ethical, political, and practical challenges of dehumanization. The following discussion looked for the points of contact and the legal options to incorporate ethical, philosophical, and political restraints. The third and fourth panels concentrated on two different manifestations of dehumanized warfare, namely, unmanned naval systems and cyber warfare. Both panels included a technical presentation to give an insight to the actual technical abilities and problems. Each panel focused on legal particularities such as the rules for precautions in attacks or the problems of distinction. These panels were able to draw a first conclusion on commonalities and a possible nexus between means and methods of dehumanized warfare. The fifth and last panel added further separate legal aspects to the discussion, such as human enhancements, the effects on the laws of war crime, the questions of humanity in warfare, and the notion of "meaningful human control," which was raised in the CCW process. This rendered the analysis more comprehensive as the focus on dehumanization was broadened so as to include additional means and methods. A central point of the conference was the conclusion based on the findings of the panels, which created the frame for the final discussion. During the discussion, the legal framework of dehumanization, the legal and nonlegal challenges of the different means and methods and possible commonalities and differences, and the nexus between law and other scientific disciplines were addressed. The goal was to produce a statement which consists of the commonly accepted principles and the challenges which have been identified in the conference. This statement is contained in this volume as a conclusion.

This would not have been possible without the work of the team of the chair of public law Professor Heintschel von Heinegg, especially Ms. Jana Wenzel, which always helped and supported us during the preparation, the conduct, and the follow-up of the conference. Also, we thank the faculty of law for its financial support and the administration of the university and its president Professor Alexander Wöll for their organizational support. Furthermore, we thank Dr. Reschke and

Ms. Schwietzer from Springer International Publishing for their extraordinary help in preparing this volume. Finally, special thanks are owed to the German Peace Foundation, which financed most of the conference, and its managing director Dr. Held, who accompanied the whole application process and always helped us in case of any difficulties.

This volume reflects the results of the conference, its discussions, and many following exchanges of thoughts and mails between all involved experts. The aforementioned final statement in the fifth panel formed the basis for the concluding part of this volume. It is addressed to scientists and students in different fields of study, practitioners, decision-makers, and experts involved in the process under the CCW.

We, the editors, hope that the content of this volume contributes to fostering further research and to enhancing the ongoing controversy concerning autonomous weapon systems and cyber warfare and is of help to anyone interested.

Frankfurt (Oder), Germany — Wolff Heintschel von Heinegg
Frankfurt (Oder), Germany — Robert Frau
Passau, Germany — Tassilo Singer
August 2017

Contents

Introduction

Wolff Heintschel von Heinegg, Robert Frau, and Tassilo Singer

Abstract Sophisticated means and new methods of warfare as well as the introduction of cyber operations pose a challenge to international law in general and humanitarian law in particular. These challenges are described in the present chapter and put under the label of "dehumanization"—a term describing the diminishing role of human beings in warfare. The chapter also gives an overview over the contents of the present volume.

A core characteristic of warfare was, always had been and is the continuous development of new methods and means of warfare to overwhelm potential enemies with new, unexpected abilities of the military or to keep at least the potential threat level against attacks at its highest possible level. Art. 36 (1) Additional Protocol I is a confirmation and recognition of that fact by law.

Since the beginning of the computer age, however, the speed of this trend accelerated. The formula of exponential growth of microchip abilities in ever smaller periods (the so called 'Moore's law') can be indirectly transferred to the increase of new military technologies, which are not solely based on better capabilities of microchips, but on multiple factors like increasingly efficient energy saving, data saving, graphics, interfaces and especially of much better software and more sophisticated program codes.[1]

[1] Compare: Advanced Copy of the Report of the 2016 Informal Meeting of Experts on Lethal Autonomous Weapon Systems (LAWS) (2016); Advanced Copy of the Report of the 2015 Informal Meeting of Experts on Lethal Autonomous Weapon Systems (LAWS) (2015), pp. 6 et seq.

W. Heintschel von Heinegg (✉) • R. Frau
Faculty of Law, European University Viadrina, Frankfurt (Oder), Germany
e-mail: heintschelvonheinegg@europa-uni.de; frau@europa-uni.de

T. Singer
Higher Court of Traunstein, Traunstein, Germany
e-mail: tassilo.singer@gmx.de

W. Heintschel von Heinegg et al. (eds.), *Dehumanization of Warfare*,
DOI 10.1007/978-3-319-67266-3_1

This enabled a rapid development of means guided by computer software, which allows to delegate more and more actions and commands to systems. As a next step increasing intra-system control abilities became possible. This also encompassed deeper intelligence, meaning the linkage of multiple problem solving (calculation) circles, enhanced situational awareness of the system itself by connecting sensors with supervision control software and real-world problem solution abilities (for example how to open doors, react to obstacles, system failures or how to handle communication break-ups).[2] All of this allows growing autonomy of unmanned systems and at the end increasing independence of distinguished computer programs for cyber operations. A final step of this development might be a fully autonomous system with high end AI (Artificial Intelligence), which is able to act unpredictably, does not need human control anymore and might actually not be supervised. Combined with now possible long endurance, high reliability, high speed, high-end sensor and communication abilities the current and coming unmanned systems can bring the efficiency of warfare to a new level.[3] The current state of the art system, the X-47 B is able to land and take off a US-carrier and could even air-refuel itself autonomously—one of the most difficult maneuvers already for human pilots.

In addition to the technical and practical advantages financial considerations endorse the development of autonomous weapon systems and cyber tools. Studies of the US-Department of Defense pointed out that the integration of preferably independent control or even autonomous abilities in unmanned systems contributes on multiple stages to cost savings, due to reduced manpower and sinking production and maintenance costs. This is of even more importance for governments in financial constraint environments.[4] Also out of purely political reasons unmanned warfare becomes increasingly attractive, especially for democratic states in guerilla wars. The participation in warfare changed to a point, wherein the majority of wars are led in non-international armed conflicts, by non-conventional means and performed by civilians. Asymmetric warfare has become the rule not the exception. Every dead soldier brought home can reduce the support for the ruling party and might result into a recall of the responsible decision makers. Thus, there is a growing reluctance of democratic states to endanger their human soldiers. However new threats are rising on every corner, particularly global terrorism, which have to be encountered and tackled somehow. Unmanned warfare seems to be an attractive solution to this "democratic dilemma".

All of these reasons combined with a changed environment on the battlefield of today, there is no doubt that the political and military leaders focus on the removal

---

[2]Compare the DARPA's Robotic Challenge and the tasks given: http://www.theroboticschallenge.org/.

[3]Compare: Advanced Copy of the Report of the 2015 Informal Meeting of Experts on Lethal Autonomous Weapon Systems (LAWS), pp. 6 f, military rationale - 8 et seq.

[4]US Department of Defense (2014), pp. V et seq., 15 et seq., 54; US Department of Defense (2012), pp. 3, 6, 19, 22, 25, 28 50; Scharre (2014), p. 12 et seq.

of human actors on the ground as far as possible from the theatre of hostilities or replace them by new methods and means of warfare.[5]

In this context unmanned systems directly contribute to that goal by replacing human soldiers and fulfilling their active role in the most dangerous parts of the battlefield. Furthermore increasing autonomy of unmanned systems can facilitate command and control, increase the speed of decision making processes by internal filtering of sensor information and thereby accelerate reactions in time-critical situations. In the end, the focus of unmanned systems and their autonomy is mainly at the replacement in dangerous and misanthropic areas and the improvement of efficiency in the battlefield. Cyber operations are used and will be used mainly for support of military operations and not as a singular tool to substitute the human in the battlefield. Cyber tools are especially directed/concentrated at creating additional means and methods of warfare, which shall directly damage the command and control capabilities of the enemy through computer networks. Their advantages over human staff are immanent and can be concluded as time efficiency, cheap costs, effectivity against armies using computer depending technologies and their possible invisibility.

However, unmanned systems, especially autonomous weapon systems, and cyber operations have in common that the human and its control are getting reduced and/or even removed out of more and more actions and decision-making processes.

As international humanitarian law is focused on regulating the conduct of hostilities, its rules had to cope with massive changes of methods and means of warfare. Besides the need for uttermost abstractness and flexibility of rules, also laws are required for an effective regulation of upcoming changes in warfare, which particularly tackle new methods and means of warfare. Especially Art. 36 AP I[6]—New Weapons—and subsidiary the more general "Martens Clause" as accepted as customary international law[7] are the expression and the result of this demand. The ICJ also confirmed in its Nuclear Weapons Advisory Opinion that even if certain weapons were not specifically addressed in the existing treaties, nevertheless "all states are bound by those rules in Additional Protocol I" as "(. . .) expression of the pre-existing customary law, such as the Martens Clause, reaffirmed in the first article of Additional Protocol I."[8] According to the mentioned Art. 1 (2) AP I civilians and combatants remain under the protection and authority of the principles of international law derived from established custom, in cases not covered by Additional Protocol 1 or by other international agreements (which could be seen

[5]Advanced Copy of the Report of the 2016 Informal Meeting of Experts on Lethal Autonomous Weapon Systems (LAWS); Advanced Copy of the Report of the 2015 Informal Meeting of Experts on Lethal Autonomous Weapon Systems (LAWS), p. 8 et seq.

[6]Protocol (No. I) Additional to the Geneva Conventions of 12 Aug. 1949 relating to the Protection of Victims of International Armed Conflicts, 8 June 1977, 1125 UNTS 3 (hereinafter AP I).

[7]ICJ, Legality of the Threat or Use of Nuclear Weapons, Advisory Opinion, I.C.J. Reports 1996, p. 226, paras. 78, 84.

[8]ICJ, Legality of the Threat or Use of Nuclear Weapons, Advisory Opinion, I.C.J. Reports 1996, p. 226, paras. 78 et seq., 84.

in new means and methods of warfare). Hence, the general rule can be deduced that no new means or method of warfare can be kept outside a framework formed by certain principles of customary international law and always has to comply with the general principles of the law of armed conflict.

Therefore, one could raise the question, whether or not the removal and/or re-placement of humans out of actions and decision making processes has legal relevance and could be utilized as a generic term for this current development, which may have common legal consequences. To describe this issue the term "dehumanization" of warfare seems most appropriate. The term "de"-humanization implies that something is or contains less human or humanity. Thereby also, the ongoing process and the increasing degree of the pushback of human control are expressed in the best way possible. In this context, the meaning of humanization should not be confused with the human habit and the moral postulate of being human and exercising humanity (as it is used in psychology). The term has to be differentiated from another, concurring use in the sense of refusing a human to be like a human and/or to be treated like a human, for example through enslaving a human or mistreatment and torture of humans.[9] Dehumanization in our understanding does not mean that a human is becoming less humane or uncaringly, but that the human itself is less in the foreground of the actual situation and is pushed more and more into the background of decision making processes in warfare. Asaro concentrates the verb "dehumanize" on the missing human judgement in the judicial process, if autonomous systems are further advanced to take targeting decisions.[10]

Further on, you can argue that the word dehumanization ends with the suffix "ization". This suffix is used to describe ongoing processes. It is not descriptive as the alternate use would mean. Instead it is a more objective, factual expression. It cannot describe a human being and human behavior properly as it reduces the human to a thing by the suffix "ization" and thereby fostering the reduction of the human itself by the choice of this wording.

The term used in our sense however expresses "the process" of "de", meaning a wording of growing distance or a dissolution of something from another thing, and humanization, meaning in its original sense of "more" human and of human control and influence over the world and its therein situated objects.

If the control of acts and decisions, which have been met originally by humans, are increasingly delegated or transferred to unmanned systems or programs, algorithms and codes, the most relevant criterion for a legal distinction becomes human control. Therefore, the most suitable description for the reduction of the human out of the battlefield and connected decision making processes is the "dehumanization of warfare".

---

[9]Compare: Haslam (2006), pp. 252–264; Kaufmann et al. (2011); Costello and Hodson (2014), pp. 175–197; Goff et al. (2008), pp. 292–306; The term is also used in this sense in international law, see: Goshray (2014), p. 355 et seq., FN 8; Shelton (2013), pp. 141–142.

[10]Asaro (2012), p. 701.

The term "dehumanization" came up before in the context of autonomous weapons. Nevertheless, there is no intense discussion about the term itself and its legal relevance in connection with human control in the development of modern technology of warfare.[11] Also, it is not used in scholarship as a generic term for all weapon systems and methods and means of warfare, which rely on reduced humans and/or human control, encompassing unmanned systems and cyber tools.[12]

However, as the law of warfare regulates the conduct of hostilities and contains special rules concerning acts of warfare, it is not impossible and unthinkable that a minimum degree of human participation in single acts of warfare might be required to fulfill all legal prerequisites for being lawful. It could be a decisive criterion for a legal analysis, though, if a human has the ability to influence, direct and control acts of an unmanned system or a cyber program as described by "dehumanization of warfare".

The factor of missing human control is meanwhile at the heart of the scholarly discussion of modern warfare, but without viewing dehumanization of warfare as a connecting element of unmanned systems and cyber warfare.[13]

Despite this change there have been no adjustments to the law of armed conflict. Since future conflicts will be distinctively shaped by the use of these methods and means of warfare, the question arises if and to what extent applicable law is adequate to determine and regulate increasingly dehumanized warfare. Simultaneously, the basic idea of an acting and responsible human combatant is shifted to a system or software that is not "liable" by law. Thus, an ethical and political debate on the necessity of human control is required. For a comprehensive approach to the question of eligibility and effectiveness of law, findings of these branches of research have to be taken into account.

Whether the current law of armed conflict is suitable to regulate autonomous weapon systems is also subject matter of the current analysis under the framework of the Certain Conventional Weapons Convention. Proponents of a total ban on autonomous systems point out that human control is a (unavoidable) premise of international law. In this context the term of "meaningful human control" has been brought on the agenda of the CCW.[14]

The outcomes of the ongoing debate on all legal aspects of UAVs (Unmanned Aerial Vehicles) can be utilized for the legal analysis of the dehumanization of warfare. Especially the legal status of UAVs and the compliance with the general

[11] Wagner (2014).

[12] Asaro (2012), pp. 687 et seq.; Wagner (Ibid.); Stroh (2014), pp. 137 et seq.; Geiss (2015), pp. 5, 23.

[13] Anderson and Waxman (2017), pp. 1101 et seq., 1109; Horowitz and Scharre (2015), Schmitt (2013a), pp. 1 et seq.; Kastan (2013), pp. 45 et seq.; Petermann and Grünwald (2011); United Nations Institute for Disarmament Research (2014), Borrmann (2014), Geiss (2015), pp. 5, 23; However arguing in this direction: Crootof (2016), p. 4.

[14] Sauer (2014), pp. 2 f.; Human Rights Watch (2012), Horowitz and Scharre (2015); Advanced Copy of the Report of the 2015 Informal Meeting of Experts on Lethal Autonomous Weapon Systems (LAWS) (2015), pp. 4 f, 11 et seq.

principles of the law of armed conflict, such as the principle of distinction, proportionality and precautions, can be considered solved (concerning most aspects) by now.[15] Especially due to the CCW process, autonomous weapon systems are also subject of the debate in international scholarship.[16] Therefore, a basis has been formed already for the analysis of unmanned, partly or fully autonomous, weapon systems and tools for cyber warfare. However, the legal implications of unmanned maritime systems have not been discussed in depth and some issues are still open for debate, such as the classification of UMS as warships according to Art. 29 UNCLOS.[17,18]

With regard to cyber warfare, the Tallinn Manual of 2013 reflects the ongoing discussion and the often controversial process to find a common approach to the legal classification of cyber operations. The Manual stresses that its purpose is directed at being a basis for further discussions and research.[19] This explains why there is a need for a second volume of the Tallinn Manual, which will be published in 2016. Several legal issues about cyber warfare are still unresolved, besides other reasons especially because of missing state practice, like attribution problems, rules for state responsibility for cyber operations conducted from within their territorial sovereignty but also particular problems like data as military objects and so on.[20]

More generally, dehumanization might pose challenges to the legal fundaments of law of armed conflict. How can autonomous weapon systems and cyber operations be subsumed precisely in the systematics of methods and means of warfare? As mentioned before, autonomous weapons and cyber tools have to be examined in accordance with Art. 36 AP I,[21] but what exact procedures are required by law—for example, a single test or continuous checks during the development or alternatively is a test of the comprehensive system/code required or is it sufficient to check

---

[15]Project of the German Peace Foundation (2010–2012), Deutsches Rotes Kreuz (2011), Petermann and Grünwald (2011), Frau (2014), Singer (2014b), Humanitär-völkerrechtliche Implikationen, pp. 55 et seq.

[16]For example: Anderson and Waxman (2017), Sassoli (2014); Advanced Copy of the Report of the 2015 Informal Meeting of Experts on Lethal Autonomous Weapon Systems (LAWS); Compare the various statements and presentations from the 2016 meeting of Experts on Laws (11.04.2016–15.04.2016), available at: http://www.unog.ch/80256EE600585943/%28httpPages%29/37D51189AC4FB6E1C1257F4D004CAFB2?OpenDocument.; Compare also FN 25.

[17]United Nations Convention on the Law of the Sea, 1833 UNTS 3.

[18]Schulz (2014), Frau (2012), pp. 84 et seq.; Kraska (2010), pp. 44 et seq.; Doswald-Beck (1995).

[19]Compare in general: Schmitt (2013a), pp. 1, 5–6; Schmitt (2017); Concerning the differing views about the Direct Participation in Hostilities: Schmitt (2013b), pp. 118 et seq.; Singer (2017), Singer (2014a), p. 22.

[20]Schmitt (2013b), Heintschel von Heinegg (2013), pp. 123 et seq.; Krieger (2012), pp. 1–21; Schmitt (2012), Schmitt (2002), pp. 365 et seq.; Singer (2014a), pp. 17 et seq.; However compare: Schmitt (2017), pp. 11 et seq.

[21]Advanced Copy of the Report of the 2015 Informal Meeting of Experts on Lethal Autonomous Weapon Systems (LAWS) (2015), pp. 6, 12 f, 16 et seq.

separate parts?[22] Furthermore dehumanized systems, methods and means have to comply with the general principles of the law of armed conflict. This particular point is one of the main battlefields of the current legal debate.[23] Nevertheless, it can be discussed if the original understanding and interpretation[24] of these rules have to be adapted to these new developments. However there might be limits to an adaption, if the rule deviates from its original purpose.[25] This might explain the demand for new rules for new methods and means of warfare or even the call for a total ban of autonomous weapons.[26] Overall, the central question of this volume is, if today's law is suitable to govern the dehumanization of warfare effectively.

From an ethical perspective it can be questioned whether or not the decision over life and death of a human can be delegated to a machine or a computer process and if autonomous weapons or code are able at all to act in an ethical correct manner.[27]

The following chapters will address all of the aforementioned issues and try to put more light into the—often emotionally and ideologically lead—debate about the legality of autonomous weapon systems, cyber warfare and thereby the dehumanization of warfare. For this purpose two guiding questions where defined in advance: Firstly, what are the possible legal challenges of the dehumanization of warfare (in the particular section) and secondly, does the relevant rule require human control? The ratio of this volume is to approach the topic of the dehumanization of warfare from different perspectives.

The first chapter provides an easy access to the topic of dehumanization of warfare and international law. It is focused on overall questions of the dehumanization and is thereby arranged more generally. It starts by a statement of Professor Dinstein on autonomous weapon systems and its legal implications. Subsequently Dr. Boothby intensely analyses Art. 36 AP I concerning the dehumanization of warfare and gives an insight in the process of weapons review.

The second chapter focuses on the ethical perspective of the dehumanization. Therein the dehumanization and the most important ethical issues which are immanent to the loss of human control are (controversially) analyzed by Professor Joerden from the perspective of legal ethics and legal philosophy.

Unmanned Systems particular at sea and land are getting addressed in the third chapter. The beginning is made by an introduction in the status quo of research and

[22] Advanced Copy of the Report of the 2015 Informal Meeting of Experts on Lethal Autonomous Weapon Systems (LAWS) (2015), pp. 16 et seq.; Sohm (2015), pp. 104 et seq.

[23] Compare concerning Art. 57 AP I: Oeter (2013), p. 183, para 449.

[24] Henckaerts (2005), ICRC (2008), Doswald-Beck (1995), FN 18; Schmitt et al. (2006); Program on humanitarian policy and conflict research (HPCR) (2010), Schmitt (2013b, 2017).

[25] Arendt (2014), p. 19 et seq.; Boothby (2011), p. 109 et seq.; Heyns (2013); Schmitt and Thurnher (2013), pp. 231 et seq.

[26] Compare: Asaro (2012), p. 687 et seq., 708; Art. 36 (NGO) (2013); Docherty (2014); Goose St (Human Rights Watch) (2014).

[27] Compare the EUI Project (2014): Autonomous Weapon Systems – Law, Ethics, Policy, Key Question 2: http://www.eui.eu/DepartmentsAndCentres/AcademyEuropeanLaw/Projects/AutonomousWeaponsSystems.aspx.

an insight in technically feasible applications. Therein Georg Heppner reports about the research concerning the autonomy of mobile robots. This is followed by an analysis of Lieutenant Colonel Thurnher concerning unmanned maritime systems and the rule for precautions in attack. Professor Heintschel von Heinegg looks for the challenges of unmanned maritime systems in international law in the last part of this chapter, however, argues against possible legal loopholes.

In contrast special issues concerning cyber warfare are discussed in the fourth chapter. To catch up with the current state of the art, as the status of the technique evolves and changes rapidly, the start is made again by an article about technical issues. One of the most challenging issues in cyber warfare is the attribution of cyber operations. Clear and convincing attribution however is needed to establish state responsibility and is thus needed for the enforcement of international law. This particular issue is discussed by Stefan Meier in the field of digital forensics, which tries to develop technical comprehensible methods for the origin of data and data streams. In the second part of this chapter Professor Stadlmeier draws the (indirect) comparison of cyber warfare to weapons of mass destruction. Therein certain rules of international law and law of armed conflict (LOAC) are transferred to cyber operations and evaluated if cyber operations pose a challenge to international law.

Finally, the fifth chapter deals with certain aspects of the dehumanization in breadth and in depth. This part contains mostly innovative issues of the dehumanization, which rarely have been evaluated from an international law perspective. Firstly, human enhancement technologies are examined by Professor Kleffner and possible challenges by international law. Also, the dehumanization of warfare raises the question of individual responsibility. If there is no direct human control, can the directing or order-issuing soldier be held responsible for acts of an autonomous system or program? Is it necessary to differentiate legally between unpredictable and predictable errors of such systems or is there a general responsibility for all defective acts? Finally, Professor Marauhn explicitly addresses an issue raised in the CCW process. It was held that there is the legal postulate of "meaningful human control" of autonomous weapon systems. However this notion of meaningful human control is analyzed and deconstructed in the light of the law of armed conflict.

At the end the core features, the central points of discussion and the commonalities of the previous articles are summarized and evaluated. Particularly, the (basic) legal framework for the dehumanization of warfare is outlined. This also includes the most relevant and controversial legal rules in this context. For all of the contested/disputed issues a solution in accordance with existing international law is presented. In conclusion, it is shown that dehumanization of warfare does not pose unsolvable legal challenges in the first instance. From an ethical perspective, however, dehumanization is highly controversial. Nevertheless it is concluded that these reservations do not find their resonance in the current law, but could be object of legal politics. Collective responsibility (state responsibility) and individual responsibility are given in any case, be it conventional warfare or be it the dehumanization of warfare. From that perspective no legal loopholes can be identified.

The volume is closed by an outlook on yet unresolved and possibly upcoming legal and ethical problems. Especially the discussion around the notion of "meaningful human control" might continue.

Concluding, this volume will provide conclusions on the ethical and legal implications of the dehumanization of warfare. It discusses the basic legal framework of autonomous weapons and cyber warfare and identifies core legal problems.

References

Advanced Copy of the Report of the 2015 Informal Meeting of Experts on Lethal Autonomous Weapon Systems (LAWS) (2015) Submitted by the Chairperson of the Informal Meeting of Experts

Advanced Copy of the Report of the 2016 Informal Meeting of Experts on Lethal Autonomous Weapon Systems (LAWS) (2016) Submitted by the Chairperson of the Informal Meeting of Experts. http://www.unog.ch/80256EDD006B8954/(httpAssets)/DDC13B243BA863E6C1257FDB00380A88/$file/ReportLAWS_2016_AdvancedVersion.pdf

Anderson K, Waxman MC (2017) Debating autonomous weapon systems, their ethics, and their regulation under international law. In: Brownsword R, Scotford E, Yeung K (eds) The Oxford handbook of law, regulation, and technology. Oxford University Press, Oxford, pp 1097–1117

Arendt R (2014) Der Einsatz autonomer Waffensysteme im Lichte des Verhältnismäßigkeits- und des Unterscheidungsgrundsatzes. In: Frau R (ed) Drohnen und das Recht. Völker- und verfassungsrechtliche Fragen automatisierter und autonomer Kriegsführung. Mohr Siebeck, Tübingen, pp 19–34

Art. 36 (NGO) (2013) "Killer Robots: UK Government Policy on Fully Autonomous Weapons. http://www.article36.org/wp-content/uploads/2013/04/Policy_Paper1.pdf

Asaro P (2012) On banning autonomous weapon systems: human rights, automation, and the dehumanization of lethal decision-making. Int Rev Red Cross 94(886):687–709

Boothby W (2011) Autonomous systems: precautions in attacks. In: Heintschel von Heinegg W (ed) International law and new weapon technologies, 34th round table on current issues of international humanitarian law. FrancoAngeli, Milan, p 109 ff

Borrmann R (2014) Autonome unbemannte bewaffnete Luftsysteme im Lichte des Rechts des internationalen bewaffneten Konflikts. Anforderungen an das Konstruktionsdesign und Einsatzbeschränkungen. Duncker & Humblot, Berlin

Costello K, Hodson G (2014) Explaining dehumanization among children: the interspecies model of prejudice. Br J Soc Psychol 53:175–197

Crootof R (2016) The meaning of 'Meaningful Human Control'. Temple Intl Comp Law J. 30 Dec 2016 draft

DARPA's Robotic Challenge. http://www.theroboticschallenge.org/

Deutsches Rotes Kreuz (ed) (2011) Humanitäres Völkerrecht Informationsschriften, Themenheft "Nicht-bemannte Waffensysteme und Humanitäres Völkerrecht", vol 24. 2/2011

Docherty B (2014) "Taking on Killer Robots," Justsecurity.org, 23 Mai 2014. Available at: https://www.justsecurity.org/10732/guest-post-killer-robots/

Doswald-Beck L (ed) (1995) San Remo Manual on international law applicable to armed conflicts at sea and explanation. Cambridge University Press, Cambridge

EUI (2014) Autonomous Weapon Systems – Law, Ethics, Policy. http://www.eui.eu/DepartmentsAndCentres/AcademyEuropeanLaw/Projects/AutonomousWeaponsSystems.aspx.

Frau R (2012) Regulatory approaches to unmanned naval systems in international law of peace and war. Humanitäres Völkerrecht Informationsschriften 2:84–91

Frau R (ed) (2014) Drohnen und das Recht. Völker- und verfassungsrechtliche Fragen automatisierter und autonomer Kriegsführung. Mohr Siebeck, Tübingen

Geiss R (2015) The international-law dimension of autonomous weapons systems. Friedrich Ebert Stiftung Study, October 2015

Goff PA, Eberhardt JL, Williams MJ, Jackson MC (2008) Not yet human: implicit knowledge, historical dehumanization, and contemporary consequences. J Pers Soc Psychol 94(2):292–306

Goshray S (2014) Targeted killing in international law: searching for rights in the shadow of 9/11. Ind Intl Comp Law Rev 24(2):355–418

Haslam N (2006) Dehumanization: an integrative review. Pers Soc Psychol Rev 10(3):252–264

Heintschel von Heinegg W (2013) Territorial sovereignty and neutrality in cyberspace. Intl Law Stud 89:123–156

Henckaerts J-M (2005) Study on customary international humanitarian law. Int Rev Red Cross 87(857):175–212

Heyns C (2013) Report of the special Rapporteur on extrajudicial or arbitrary executions, Christof Heyns. United Nations, General Assembly, Human Rights Council, A/HRC/23/47

Horowitz M, Scharre P (2015) Meaningful human control in weapon systems: a primer. CNAS Working Paper March 2015

Human Rights Watch (2012) Losing humanity: the case against killer robots

ICRC (2008) Interpretive guidance on the notion of direct participation in hostilities under international humanitarian law. Int Rev Red Cross 90(872):991–1047

Kastan B (2013) Autonomous weapon systems - a coming legal "Singularity". J Law Technol Policy 45(2013):45–82

Kaufmann P, Kuch H, Neuhäuser C, Webster E (eds) (2011) Humiliation, degradation, dehumanization – human dignity violated. Library of ethics and applied philosophy, vol 24. Springer International

Kraska J (2010) The law of unmanned naval systems in war and peace. J Ocean Technol 5:43–68

Krieger H (2012) Krieg gegen Anonymous – Völkerrechtliche Regelungsmöglichkeiten bei unsicherer Zurechnung im Cyberwar. Archiv des Völkerrechts, 1–21

Oeter S (2013) Methods and means of combat. In: Fleck D (ed) Handbook of international humanitarian law, 3rd edn. Oxford University Press, pp 115–230

Petermann T, Grünwald R (2011) Stand und Perspektiven der militärischen Nutzung unbemannter Systeme. Büro für Technikfolgenabschätzung beim Deutschen Bundestag, Arbeitsbericht Nr. 144, Mai 2011

Program on humanitarian policy and conflict research (HPCR) (2010) The HPCR manual on international law applicable to air and missile warfare (AMW Manual). Harvard University, Cambridge University Press, Cambridge

Project of the German Peace Foundation, Völkerrechtliche Einhegung unbemannter bewaffneter Systeme, Prof. Dr. Thilo Marauhn, Mai 2010 – Juni 2012. http://www.bundesstiftung-friedens forschung.de/index.php/aktuelle-foerderung/17-details-foerderprofil/130-marauhn

Sassoli M (2014) Autonomous weapons and international humanitarian law: advantages, open technical questions and legal issues to be clarified. Intl Law Stud 90:308–340

Sauer F (2014) Autonome Waffensysteme. Humanisierung oder Entmenschlichung des Krieges? Glob Gov Spotlight 4:2–4

Scharre P (2014) Robotics on the Battlefield Part I, Range, Persistence and Daring. Center for New American Security, 20YY Series

Schmitt M (2002) Wired warfare: computer network attack and Ius in Bello. Int Rev Red Cross 84(846):365–399

Schmitt M (2012) International law in cyberspace: the Koh speech and Tallinn Manual Juxtaposed. Harv Int Law J 54:13–37

Schmitt M (2013a) Autonomous weapon systems and international humanitarian law: a reply to the critics. Harv Natl Secur J. Online Content, Features, 05.02.2013. http://harvardnsj.org/2013/02/autonomous-weapon-systems-and-international-humanitarian-law-a-reply-to-the-critics/

Schmitt M (ed) (2013b) Tallinn Manual on the international law applicable to cyberwarfare. Cambridge University Press

Schmitt M (ed) (2017) Tallinn Manual 2.0 on the international law applicable to cyber operations. Cambridge University Press

Schmitt M, Thurnher J (2013) "Out of the Loop": autonomous weapon systems and the law of armed conflict. Harv Natl Secur J 4:231–281

Schmitt M, Garraway C, Dinstein Y (2006) The Manual on the law of non-international armed conflict, with commentary. International Institute of Humanitarian Law, Sanremo

Schulz M (2014) Autonomie zur See – Die völkerrechtliche Einordnung von unbemannten militärischen Seefahrzeugen. In: Frau R (ed) Drohnen und das Recht. Völker- und verfassungsrechtliche Fragen automatisierter und autonomer Kriegsführung. Mohr Siebeck, Tübingen, pp 103–118

Shelton D (ed) (2013) The Oxford handbook of international human rights law. Oxford University Press

Singer T (2014a) Cyberwarfare – Damoklesschwert für das Völkerrecht?. Sicherheit und Frieden 1:17–23

Singer T (2014b) Humanitär-völkerrechtliche Implikationen der Drohnenkriegführung. In: Benedek W, Folz HP, Isak H, Kettemann MC, Kicker R (eds) Bestand und Wandel des Völkerrechts, Beiträge zum 38. Österreichischen Völkerrechtstag 2013 in Stadtschlaining. Peter Lang, Frankfurt a.M., pp 55–73

Singer T (2017) Update to revolving Door 2.0 – the extension of the period for direct participation in hostilities due to autonomous cyber weapons. In: Roigas H, Jakschis R, Lindström L, Minarik T (eds) Defending the Core, 9th international conference on cyber conflict 2017. NATO CCDCOE Publications, pp 121–133

Sohm S (2015) Völkerrechtliche Pflichten im Vorfeld militärischer Gewaltanwendung – Aktuelle Entwicklungen bei der Prüfung neuer Waffen und Methoden der Kriegführung. J Int Law Peace Armed Confl 3:104–110

Steve Goose (Human Rights Watch) (2014) Statement to the Convention on Conventional Weapons Informal Meeting of Experts on Lethal Autonomous Weapons Systems. https://www.hrw.org/news/2014/05/13/statement-convention-conventional-weapons-informal-meeting-experts-lethal-autonomous

Stroh P (2014) Das Menschenrecht auf Leben im zunehmend "entmenschlichten" bewaffneten Konflikt. In: Frau R (ed) Drohnen und das Recht. Völker- und verfassungsrechtliche Fragen automatisierter und autonomer Kriegsführung. Mohr Siebeck, Tübingen, pp 137–162

United Nations Institute for Disarmament Research (2014) The Weaponization of Increasingly Autonomous Technologies: Considering how Meaningful Human Control might move the discussion forward, no. 2 UNIDIR Resources

US Department of Defense (2014) Unmanned Systems Integrated Roadmap. FY 2013–2038, 14-S-0553

US Department of Defense, Defense Science Board (2012) The Role of Autonomy in DoD Systems, Task Force Report. July 2012

Wagner M (2014) The dehumanization of international humanitarian law: legal, ethical, and political implications of autonomous weapon. Vanderbilt J Transnational Law, 47. University of Miami Legal Studies Research Paper No. 15-1

Part I
Dehumanization of Warfare and International Humanitarian Law

Autonomous Weapons and International Humanitarian Law

Yoram Dinstein

1. Ever since the advent of the industrial revolution in the nineteenth century, weaponization in warfare has been undergoing a technological revolution of its own. The radical transformation in the field has rapidly accelerated in the era of computers and electronics. Some modernistic weapons—such as blinding laser beams—have been spurned by States and banned by treaty (usually on the ground that they cause superfluous injury or unnecessary suffering). But the vast majority of state-of-the-art means of warfare have met little or no resistance by the international community, and have been smoothly incorporated in the arsenals of actual or potential Belligerent Parties eager to employ them and able to afford the expense entailed.
2. When novel weapons catch the attention of the public at large, Governments—particularly in democratic countries—cannot be entirely oblivious to popular perceptions and even prejudices. Whereas public opinion may occasionally be mesmerized by a pioneering technology facilitating the production of new weapons, more often than not it is predominantly swayed by the injurious effects of their use on eventual victims: combatants and, even more so, non-combatants. It is no accident that outdated anti-personnel landmines have aroused much more passionate debate (ultimately forcing most Governments to relinquish their use) compared to, say, the introduction of cutting-edge stealth aircraft. However, it is noteworthy that one particular aspect of current weaponization is hard for the "civil society" to digest. That is the phenomenon of automatization in air warfare,

Y. Dinstein, Prof. Em. Tel Aviv University; Member, Institut de Droit International.

Y. Dinstein (✉)
Tel Aviv University, Tel Aviv, Israel
e-mail: ydinstein@gmail.com

W. Heintschel von Heinegg et al. (eds.), *Dehumanization of Warfare*,
DOI 10.1007/978-3-319-67266-3_2

namely, the displacement of flesh-and-blood pilots from the cockpits of military aircraft wherein human presence was previously taken for granted.

3. The layman's disposition to believe that a pilot should be ensconced on board flying platforms affects the current public discourse concerning drones (viz. unmanned aerial vehicles). The launch of missiles or "smart" bombs (equipped with precision guidance) from manned military aircraft has been largely taken in stride by public opinion, subject to justifiable concerns about their potential margin of error in targeting. Yet, when the same missiles or "smart" bombs are unleashed by unmanned drones, there is an outpouring of atavistic fears about an unblinking evil eye in the sky. Rationally, drones must be recognized for what they are, i.e. remotely-controlled flying platforms (which may or may not be armed). The fact that drones are unmanned is a testament to technological sophistication in design, but there is no reason for treating them as a sinister quantum leap in weaponization.
4. In operational terms, weapons and weapon systems—manned or un-manned—are frequently integrated, and the introduction of remote control affects not only drones. This is illustrated in two disparate scenarios:

   (a) An unarmed drone may be instrumental in acquiring a target, but the latter may then be struck by a missile launched from a manned warplane available in the battle space.
   (b) A manned military aircraft may be driven to a given coordinate where it releases a "smart" bomb, which may then be taken over and directed to target by fire controllers who are not on board.
   In both instances, flight by a platform (whether manned or unmanned) does not necessarily coincide with the delivery of a payload to the target: synergy is required between two different players.

5. Drones are merely a species of a new genus of weapon systems (not restricted to air warfare) endowed with some autonomy in operation. The use of such devices is growing unabated, commensurately with relentless breakthroughs in high technology. All the same, it must be appreciated that new weapon systems diverge in many respects not only from older means of warfare but also from one another. Each novel weapon system has its own paradigmatic characteristics and what counts is function rather than form.
6. Drones are multi-tasked in that they can:

   (a) Fly entirely on auto-pilot.
   (b) When unarmed and used exclusively for surveillance, be miniaturized in size (thus reducing the risk of being shot down).
   (c) Carry out the crucial mission of surveillance of enemy targets in real time, collecting information through the use of cameras and various sensors (infra-red, electro-magnetic, thermal, and the like). This is indispensable for evaluation of the situational canvas prior to embarking on an attack.

(d) Loiter over and around a given military objective for many hours, securing sequential and multi-layered data by intermittently changing altitudes and angles of reconnaissance.
(e) Fly at considerable heights (above the ceiling of many air defence systems and interceptor aircraft) and obtain information even through thick cloud layers.
(f) Hibernate and then reawaken to continue the assignment at different time-frames.
(g) Move from one target to another without necessarily having to return to base first.
(h) When armed, launch missiles or drop "smart" bombs, enabling "on-the-nose" deadly strikes that may take the enemy by complete surprise.
(i) When unarmed, direct to target the payload carried on other (inter-linked) platforms.
(j) Whether a mission is successfully accomplished or aborted, land safely with all the expensive electronic equipment on board. This is in contrast to cruise missiles, which are designed for a single use (occasionally putting in doubt their cost-effectiveness).

All these are only highlights of the drones' capabilities.

7. Unlike drones, contemporary land robots may be quite narrow-gauged in their operational spectrum. Although self-propelled, their employment may be limited to well-defined non-versatile purposes, focusing for instance on one of the following missions:

(a) Patrol along an exposed road.
(b) Detection of booby-traps or IEDs (improvised explosive devices).
(c) Provision of supplies to front-line units.
(d) Collection of wounded from the field.

8. Maritime robots may have even more sharply-defined aims. Innovative naval mines—fitted with modulated sensors (based on acoustics, water pressure, etc.)—can be placed at the bottom of the sea and programmed to be activated by the signature of preselected types of surface warships or submerged submarines, to the exclusion of other vessels. When a target is verified, the naval mine transforms itself into a torpedo: it then chases and seeks to destroy the prey.
9. Whether hovering in the air, moving on land or floating in water, military robots are all implements which—to a greater or lesser degree—may be labeled as autonomous weapons or weapon systems. The autonomy in question relates to attack capability and not just to an auto-pilot or self-propelled proficiency. The distinguishing feature of an autonomous weapon system is that, upon activation, it can strike targets without further intervention by a human operator. In other words, an autonomous weapon system is able to engage targets on its own—relying on preprogramming or on-board sensors—without human guidance subsequent to activation.

10. Overall, there are three diverse types of advanced technology weapon systems available today. These systems may be classified as (a) partially-autonomous, (b) semi-autonomous, or (c) fully-autonomous. The distinction between the three categories is predicated on the extent of human involvement in mounting an attack once the data input is processed.
11. The first type is that of "man-in-the-loop". Here a human being is entrusted with operating an unmanned device (epitomized by an armed drone) by remote control. The person in charge may filter and appraise the information streaming in from the cameras and the sensors of the drone. He also directs the mission accomplishment of the drone, especially by finalizing the execution of an attack against a specific objective or deciding to refrain from carrying it out. Clearly, this type of technology—albeit advanced—is by no means genuinely autonomous. The man is in overall control, even if he sits next to a screen located far away and manipulates the machine by punching instructions on a keyboard. As for the term "man", it must be understood as transcending gender. Increasingly (especially in Israel), the responsibility is assigned to a woman.
12. The second type of advanced technology weapon system is that of "man-on-the-loop", where the artificial contrivance can use on-board software to process data and engage enemy targets—without additional human input subsequent to activation—but a human overseer can still monitor its functioning. In tracking progress, that person retains the power to intervene and override the configurations of the apparatus for whatever reason (primarily, if the attack under way is likely to run afoul of the international legal precepts of armed conflict: governing distinction and proportionality). This category of weapon system is semi-autonomous.
13. The "man-in-the-loop" or the "man-on-the-loop" may be far removed physically from the strike zone, running or monitoring the mechanical rigging from across a continent or an ocean. In fact, the person may even be a civilian (this would be the case, for example, when a CIA agent—who is not a member of the US armed forces—aims armed drones at targets in Yemen or Pakistan). But, if so, we are not talking about an ordinary ("innocent") civilian. The person in or on the loop will be considered a civilian directly participating in hostilities. The rule is that, for such time as a civilian is directly participating in hostilities, he is assimilated by the law of armed conflict to a combatant and therefore loses civilian exemption from attack.
14. As for members of the armed forces operating a drone, whether in or on the loop, they are obviously combatants. Still, they may be deemed unlawful combatants should they be using for belligerent purposes a civilian drone (i.e. a drone not bearing military markings). If captured by the enemy, unlawful combatants would be deprived of entitlement to the privileged status of prisoners of war (which is bestowed only on those meeting certain conditions set forth by the law of armed conflict).
15. The third type of advanced technology weapon systems is that of "man-out-of-the-loop". Here the device used has a fully autonomous property in that, following activation, it is capable of acting entirely on its own (pursuant to

preprogrammed parameters) without any override capability being reserved following human review. In excluding further operator's guidance or involvement in the aftermath of a launch, the mechanism has a "deploy-and-forget" attribute indicating that it is capable of selecting by itself the best mode of mission accomplishment (as defined earlier by its human masters). Contemporary usage is admittedly circumscribed in scope. Chiefly, fully autonomous weapons are designed for radar-jamming or analogous, rather modest, purposes.

16. All three subdivisions of existing autonomous weapon systems are confined to built-in parameters constraining their activities. But the buzz nowadays is about a whole new technology in the making, to be probably ushered in approximately two decades from this moment in time. The "next generation" weaponry, we are told by scientists and armaments' specialists, will feature robots with AI (artificial intelligence) who will be capable of improvising in the battlefield—making their own judgment calls in the face of unforeseen events—while free in every respect from human intervention.
17. Some elements of AI technology have already gained access to current robot development. By now, some land robots can learn from their own previous errors in, e.g., surmounting or avoiding physical obstacles impeding the traversing of terrain. But these are relatively narrow skills. The challenge for future robots is to advance from physical to intellectual hurdles, exercising discretion when encountering complex scenarios.
18. The emergence of profoundly cognitive robots, if and when it materializes, is bound to alter the face of warfare. But will AI robots be able to perform better than humans? On the one hand, as machines, AI robots will operate without being prone to human errors that detrimentally affect so many military decisions under stress today. On the other hand, there is the counterpart possibility (statistically far from negligible) that, like all machines, AI robots will be plagued with technical malfunctions even beyond the early stage of "cutting their teeth".
19. It has been contended that AI robots may have the edge over any human rivals. They will naturally be vested with enhanced vision, superior strength and augmented speed; they will benefit from greater capabilities and skills; and they will possess extended endurance power. But, above all, their judgment will not be clouded by emotions. Some enthusiasts even rave about AI robots fundamentally proving to be more humane than humans. Nevertheless, a total lack of any emotions is liable to become a downside rather than an upside. This is underscored by an absence of compassion towards random victims of war who are exceptionally vulnerable (like children or persons with disabilities).
20. The principal problem that the law of armed conflict will have to address upon the baptism by fire of AI robots relates to the application and calculation of proportionality in attack. It is not apparent at the present juncture how machines can be devised in the incredibly complex and nuanced fashion that will enable them to be capable—without being assisted by human beings—of evaluating in concrete circumstances whether collateral damage to civilians/civilian objects (incidental to an attack against lawful targets) is expected to be "excessive"

compared to an anticipated military advantage. Human beings are often frustrated in attempts to figure out whether collateral damage to civilians/civilian objects is "excessive" in which case an attack against a lawful target has to be aborted. How can it be envisaged that AI robots will be up to the task?

21. There are abundant additional interlinked issues:

    (a) Will future AI robots be solely military in nature or will technology craft—side by side with them—civilian AI robots, dedicated to the discharge of humanitarian duties in the course of hostilities (whether in the medical treatment of the wounded and sick, in the domain civil defence or in related spheres of activities)?
    (b) If so, how will the two separate sub-sets of AI robots be told apart? Presumably, military robots will necessarily have to be marked by distinctive military insignia.
    (c) Then, what will ensue if a military AI robot (acting on its own initiative) decides to divest itself of the telling military marking?
    (d) Above all, what will happen if a military AI robot were to perfidiously feign civilian status in order to kill or injure enemy personnel?

22. The crux of the issue is that, by definition, AI robots will be unpredictable. What will happen if an AI robot chooses to go rogue (notwithstanding prescribed parameters), by turning against all human beings or against some of them? And what will transpire if an AI robot opts to desert to the enemy? The main dilemma, perhaps, will relate to the application of international criminal law. After all, how will it be possible to punish an AI robot for a war crime? Would the offensive AI robot be destroyed (deactivated)? Would those in charge be required to resort to any "due process" prior to taking such drastic measures against the AI robot?
23. A supplementary question relates to the residual criminal accountability of any human beings for the misdeeds of AI robots who act in breach of the law of armed conflict (or even domestic law). It is imperative to bear in mind that AI robots—like all weapon systems—will still be conceived and activated by human beings. In principle, robotics engineers may be held individually accountable (in a criminal extension of the legal concept of product liability) if their creatures do not comply with the legal norms of armed conflict. Military commanders may also bear responsibility for the actions of AI robots that they let loose. However, what is the scale of penal blame that can be pinned on human beings if AI robots inscrutably defy unequivocal instructions encoded in their electronic circuits?
24. I am sure that numerous other conundrums will come to light once AI robots proceed from the drawing board or the laboratory phase to the arena of actual "weapon review" (a mandatory procedure for prospective means of warfare pursuant to the law of armed conflict) and, eventually, the empirical realm of practical application. Long-range legal conjecture can go only so far, and in the final analysis there will have to be a reality check. A new slate of existing conditions will surely trump all advance speculation.

Dehumanization: Is There a Legal Problem Under Article 36?

William Boothby

Abstract While remote attack, whether using remote piloting, autonomous attack technology or cyber techniques, does not per se raise legal issues, there is a clear ethical dimension. People are nevertheless closely involved, fulfilling various critical roles. All forms of mechanical learning associated with attack technologies are not unacceptable. Consider, for example, learning methods integrated into a weapon system that are designed to increase or ensure victim protection. Ethical concerns will, however, persist and will be associated with concerns that machines should not be permitted to decide who is to be attacked and who is to be spared. Zero casualty warfare is not as such unlawful. Customary and treaty rules of weapons law apply to these weapon technologies including the obligation for states to undertake weapon reviews. The Chapter summarises these customary and treaty rules and notes that reviewing autonomous weapon technologies will involve an assessment of whether the weapon system is capable of undertaking the decision-making that targeting law requires, and to which reference is made in the Chapter.

1 Introduction

At long last, the idea that States are legally obliged to undertake weapon reviews is attracting attention. For most states in the world,[1] the obligation is set forth in article 36 of the First Protocol Additional to the 1949 Geneva Conventions (API) in the following terms:

> In the study, development, acquisition or adoption of a new weapon, means or method of warfare, a High Contracting Party is under an obligation to determine whether its

W. Boothby, Associate Fellow at the Geneva Centre for Security Policy.

[1]At the time of writing, there are 174 states party to the Protocol Additional to the Geneva Conventions of 12 August 1949, and Relating to the Protection of Victims of International Armed Conflicts, Geneva, 8 June 1977 (API), www.icrc.org viewed on 21 January 2015.

W. Boothby (✉)
Geneva Centre for Security Policy, Geneva, Switzerland
e-mail: williamboothby@hotmail.com

W. Heintschel von Heinegg et al. (eds.), *Dehumanization of Warfare*,
DOI 10.1007/978-3-319-67266-3_3

> employment would, in some or all circumstances, be prohibited by th[e] Protocol or by any other rule of international law applicable to the High Contracting Party.

For States that are not party to API, there is a implied obligation legally to review new weapons,[2] but no customary or implied duty legally to review new methods of warfare. However, all methods of warfare adopted by States must be in compliance with international law and States will therefore wish to evaluate them in advance to ensure that this is so. So what exactly do the terms 'weapon', 'means of warfare' and 'method of warfare' mean?

A weapon comprises an offensive capability that can be applied to a military object or enemy combatant.[3] The means whereby this is achieved will involve a device, munition, implement, substance, object, or piece of equipment that is used, that it is intended to use or that has been designed for use to apply the offensive capability, usually causing injury or damage to an adverse party to an armed conflict.[4] It is that device, etc. that is generally referred to as a weapon. The phrase 'means of warfare' refers to all weapons, weapons platforms, and associated equipment used directly to deliver force during hostilities. Methods of warfare consist of the general categories of operations, such as bombardment or blockade, and can refer to particular tactics or ways of conducting hostilities.[5] It follows from this that ways of undertaking hostilities in which the involvement of people is diminished must, as with other new methods of warfare, be considered for legal compliance in accordance with article 36 by States party to API.

It should, as a preliminary matter, be noted that while the use of new types of weapon and new methods of undertaking hostilities may cause human involvement to be reduced or diminished they will not cause that involvement to be removed entirely. Thus, cyber warfare methods, for example, may enable damage and/or injury to be caused to an adversary without a member of the attacking force needing to be present at or near the location where the injurious effects occur. Remotely piloted aircraft (RPA) may cause injury or damage but the operator may remain at a very considerable distance from the scene of the destruction. Autonomous platforms may undertake attacks without the persons who authorised or initiated the relevant sorties being anywhere near the place where the operational effect

---

[2]Consider, for example, Common Article 1 to the Geneva Conventions, Regulations Respecting the Laws and Customs of War on Land, annexed to Hague Convention IV, The Hague, 18 October 1907 (Hague Regulations), article 1, HPCR Manual on International Law Applicable to Air and Missile Warfare (AMW Manual), Rule 9 and Tallinn Manual on the International Law Applicable to Cyber Warfare [Tallinn Manual (Schmitt 2013)], Rule 48a.

[3]McClelland (2003), p. 397.

[4]The HPCR Manual on the International Law Applicable to Air and Missile Warfare (2010) (AMW Manual) defines a weapon as 'a means of warfare used in combat operations that is capable of causing either (i) injury to, or death of, persons; or (ii) damage to, or destruction of, objects'; AMW Manual, Rule 1(ff). Munitions, or ammunition, constitute a narrower concept referring to the object that actually causes the injury, death, damage or destruction. AMW Manual, Commentary accompanying Rule 1(ff), para. 4 cites bombs or missiles as examples of munitions.

[5]See for example AMW Manual, Rule 1v.

takes place. However, while such methods may appropriately be characterised as involving 'remote attacks', that element of remoteness does not mean that people are uninvolved in such operations. Quite the reverse is true. People are and will remain of fundamental importance in the conduct of such hostilities. It will be the decision of a human Commander that will authorise the conduct of the cyber, RPA or autonomous mission. Scientists, development engineers and computer programmers will write the computer code that enables the cyber attack to be undertaken, will design and construct the RPA that is used to attack a target and will create the artificial intelligence on which the operation of the autonomous weapon system is based. Technicians will install the mission critical data, will have planned the sortie or operation, will load the ammunition and fuel, will have serviced the airframe or other platform, and will maintain the computing equipment that is used to undertake the cyber attack. Human beings will load relevant data into target recognition software, and it will be people who will evaluate the success or otherwise of the mission and who will adjust future operations accordingly. While this is not, and is not intended to be, a complete list of categories of human involvement in such operations, it does illustrate the falsity of suggestions that such technologies will preclude human involvement in hostilities.

Nevertheless, it is proper to point out that the nature of human involvement in 'remote attack' does differ from what we see in more conventional methods of warfare. While people in the vicinity of the locus of the attack remain at risk, sometimes severe risk, of injury or death from remote attack, the attacker is operating from what will frequently appear to be a 'safe distance'. While this may appear to some observers to be unethical, perhaps because it offends some notions of chivalry, to others it merely represents the most recent manifestation of a tendency started when the ancient Greeks developed early forms of bow, of cannon and of crossbows. Indeed, maybe our contemporary concerns about remote attack have their roots in the Homeric complaint that such tactics of undertaking attacks from a distance are not heroic.[6]

If we are sensibly to analyse whether these novel methods of warfare comply with applicable rules of international law, we need to consider a little more closely what the new methods might involve. RPA have been in frequent use since 2002, for example by the US in Afghanistan and Pakistan and by the UK in Afghanistan, to conduct air attacks against ground targets, predominantly using Hellfire missiles. Numerous other States are known to have RPA progammes and some have used such technologies for ground attack. Remotely controlled weapon systems are not, however, limited to the air domain. Cyber methods have already been employed as a way of causing damage to a State's assets. The Stuxnet attack on the Iranian nuclear centrifuges, discovered in 2010, is by no means the only example of a cyber operation that targets the property and infrastructure

---

[6]Idomeneus, referring to the bow, complained "my way is not to fight my battles standing far away from my enemies"; Homer, Illiad, 13.262-3. The use of the bow was not consistent with the confrontational image that was the essence of heroic warfare; O'Connell (1989), p. 48.

of a State. Computer defences against such attacks are classed by the Tallinn Manual in terms of passive and active cyber defence[7] from which it can be reliably deduced that active cyber defence measures may involve launching pre-emptive or preventive cyber counter-operations against the source of the initiating attack. Some States are developing missile defence systems that, when active, detect inbound missiles and respond automatically by firing a munition to intercept the inbound missile.[8] Research focuses on the development of autonomous attack technology that would be capable of use on unmanned platforms. Technicians would, for example, develop the software and install algorithms corresponding to a chosen type of target; the platform would move to the target area where it would employ its sensors to search an area which might or might not have been defined by pre-set co-ordinates. The search might be limited to a specified time period and involve looking for targets that match the algorithmic data. If such an object is found, it would be the machine that then decides when and how to undertake the attack.

Autonomy and significant levels of automation may also be employed in connection with cyber operations. Cyber attacks in defence are likely to be automatic, pre-programmed cyber operations against intrusions from the outside, sometimes referred to as 'hack-backs'. Given that the initiating attack may have come from, or via, a multiplicity of computers which are not necessarily military in nature and which may not necessarily be operated by a party to the conflict or indeed have any other involvement in the conflict, "states will have to carefully evaluate the lawfulness of such automatic hack-backs in light of the principle of precaution."[9]

So while some elements of remote attack technologies are already in service others lie at or beyond the horizon. These advances in technological prowess should not however blind us to the fact that many armed conflicts will for the foreseeable future inevitably continue to be fought using more traditional, manned operations in which people will continue to be more directly and centrally involved in the application of force, including by hand to hand fighting. The existing body of law, as we shall see, applies across this increasingly broad spectrum.

---

[7] In the Tallinn Manual Glossary 'active cyber defence' is defined as "A proactive measure for detecting or obtaining information as to a cyber intrusion, cyber attack, or impending cyber operation, or for determining the origin of an operation that involves launching a preemptive, preventive or cyber counter-operation against the source." 'Passive Cyber Defence' is defined as "A measure for detecting and mitigating cyber intrusions and the effects of cyber attacks that does not involve launching a preventive, preemptive or countering operation against the source."

[8] See for example Israel stages test flight of Arrow 3 missile defence, Daily Telegraph, 25 February 2013 available at http://www.telegraph.co.uk/news/worldnews/middleeast/israel/9893150/Israel-stages-test-flight-of-Arrow-3-missile-defence.html.

[9] Droege (2012), p. 574 and for a discussion of the precautions principle as it applies to cyber warfare, see ibid, pp. 575–576.

2 Does the Depersonalisation of Warfare per se Raise Legal Issues?

As the trends in technological development mentioned in the previous section of this paper place attackers at ever greater physical distances from the scene of the violence that they unleash, the first question that arises is whether remote attack per se raises legal issues. The short answer is 'no'. There is no legal objection to attacking the enemy from a safe distance. Specifically, there is no legal requirement to accept any risk to one's own forces as some kind of pre-condition to the lawful application of force against the enemy. Indeed, it is one of the more fundamental aims of military strategists to seek to maintain the integrity of own forces while exposing the enemy to destructive fires, and remaining 'beyond arm's length' has always been one favoured method of achieving this. The technologies we have been discussing do of course extend the relevant distances dramatically. It is, nevertheless, clear that remoteness in attack and the consequent depersonalisation of the battlefield do not per se constitute a legal issue to be addressed, for example, during a legal review of such a method of warfare.

3 The Ethical/Legal Dimension

If international law concerns are not aroused by the mere remoteness as such of an attack method, do some such technical developments raise ethical issues? Put somewhat simplistically, the question here seems to be whether it is ethically acceptable for weapon systems to decide who is to live and who is to die, what is to be damaged and what is to be spared. Weapon systems, after all, are not sentient beings, they are not endowed with emotions, with the capability to exercise judgment, to experience empathy and compassion and are not liable to the criminal and disciplinary sanctions confronting a combatant. These objections would, for some, render autonomous attack technology fundamentally objectionable for the indefinite future.[10]

We have already noted that people are always going to be most closely involved even in autonomous attacks. By designing the software that is used to perform the autonomous targeting decision making, by prescribing the area that the sensors must search and the time limitations of such a search on the occasion of a particular sortie, by setting the number of points of similarity that there must be for acceptable recognition of a lawful target to be deemed to have occurred, by reviewing the pattern of life data when planning the sortie and by deciding whether to allow the sortie to take place human beings set close constraints on what the autonomous weapon system can do. 'Autonomy' in this sense is always going to be confined

[10]Consider, for example, Sparrow (2007), from p. 66.

by pre-determined boundaries. It follows that such an autonomous weapon system is and remains the tool of the human being who authorizes the mission. An RPA is even more directly the tool of the Commander who authorises the mission and of the operator who controls it. Similarly, a cyber weapon is the tool, for these purposes, of those who command, direct and initiate the cyber operation in which it is employed. It is that aspect which seems to be critical both to the ethical and to the legal issues that such operations raise.

If a weapon system is pre-set to react in a particular, hostile way if specified events occur, for example when the Iron Dome system engages inbound rockets, the nature and circumstances of this permitted response have been decided and then programmed by human beings. In such circumstances, the weapon system remains the tool of the individual who authorizes its deployment on such a mission. If the status of the weapon system as the 'tool of the individual authorising the mission' is as critical as the author believes it to be, the question arises when does the system cease to have that status. Views will differ,[11] but it is arguable that that point of transition arise when the weapon system applies learned lessons to go beyond constraints imposed by those initiating the mission, or when it is given complete discretion as to what is to be attacked, when, where, in what circumstances and how. One can distinguish those situations from circumstances in which the weapon system declines to attack because of something its sensors have observed in the vicinity of the target, or when it makes decisions the parameters of which have been carefully prescribed in advance by those that develop and employ the platform, instrument or cyber tool. There would seem to be no legal or ethical concerns if the 'self-initiated' decisions that the weapon system is making amount to application to itself of additional constraints limiting attacks to those that are legally permissible, for example by detecting whether any unexpected persons or objects are detected in the vicinity of a target. Ethical and legal concern is likely to arise when the weapon system applies artificial intelligence to liberalise the circumstances in which it can use force or to make its own decision as to the designation of an object or person as a target to be attacked.

Clearly, other observers will regard the critical factor as the making of a tactical attack decision, concluding that any platform or system that is designed to decide mechanically or automatically what to attack, when and perhaps how, raises ethical concerns. For the author, the distinction that matters is whether the weapon system is operating according to pre-set constraints or is operating in an unconstrained way, either because constraints have not been applied to the mission or because the weapon system's controlling logic permits it to go beyond the constraints that were initially programmed.

[11]Steven Haines makes the point that war is always a human, moral and social activity and that "a conflict 'fought' exclusively by machines against other machines could not constitute a Clausewitzian war". Haines believes it to be part of the nature of war that all those engaged in it must be capable of reaching moral judgments as to what they are doing; Haines (2012), pp. 9 and 11.

It would be wrong to conclude, however, that all forms of mechanical learning by weapon systems are going to be unacceptable. Consider, for example, an unmanned aircraft equipped with autonomous attack facilities searching an area that has been defined by the person who planned the sortie. It looks for objects that comply with points of recognition prescribed by the planner, comparing what it observes in the area surrounding the target with the data fed into it by that individual, a comparison process that was informed by pattern of life observations made during the period leading up to the sortie. If additional artificial intelligence were to be applied to such a system, with the consequence that the system learns how to detect protected persons or objects such as civilians or civilian objects more reliably, this would seem to be acceptable both legally and ethically.

By contrast, an artificial intelligence system that permits the weapon system to loosen pre-set constraints that reflect targeting law will of course be legally and presumably ethically unacceptable. Between these relative extremes, testing should be employed to determine exactly how the weapon system's learning process is limited and whether the results of all possible learning outcomes will be such that the platform's decisions, and the precautions taken in reaching them, comply with targeting law rules.

A fundamental, philosophical[12] question which goes to the root of the acceptability or otherwise of all forms of remote attack is whether there is some sort of 'participation fee' that must be paid in order to justify involvement in an armed conflict and whether that 'fee' is expressed in blood? Perhaps this is the thought that lies at the root of the unease we may feel about depersonalized warfare. Or perhaps we feel that people must remain involved 'at the sharp end' as it were, because only by being so can people instantly decide when the war should end. Or maybe we just feel that war is, and should remain, at core a human activity in which people apply force more or less directly and immediately. After all, it will have been disputes among humans that initiated the war and the outcome of a war fought between machines may prove nothing other than which party to the conflict is the technically superior party, with the implicit objection that warfare is about identifying something different, though exactly what may be hard to define. There is the further objection that establishment of technological superiority could be achieved without any use of weapons or methods of warfare and without the infliction of any casualties or damage.

A legal analysis of these issues should conclude that resort to the use of force does not necessarily imply an obligation to pay any 'fee', whether expressed in blood or otherwise. A State is only permitted to resort to the use of force in exercise of its inherent right of individual or collective self-defence[13] or as authorized by

[12]'Philosophical' is not used here in a technical sense.

[13]UN Charter, article 51.

a resolution of the UN Security Council under Chapter VII of the UN Charter.[14] If a State is attacked by another state or if it is operating to restore international peace and security, it is difficult to see what good ethical or philosophical reason there might be why it should not use purely autonomous or other remote attack methods, alone or in combination with other methods, to exercise that right of self-defence or to restore peace and security. Put another way, there would seem to be no legal, ethical or philosophical basis for saying that a State defending itself against unlawful attack or undertaking Security Council-mandated action must place its own personnel, civilian or military, at any risk at all in doing so.

Specifically, the law of armed conflict does not prohibit 'zero casualty' warfare in which one side seeks to fight without suffering casualties, provided that implementing such a policy does not involve a failure to implement the law of targeting rules, including the rules as to precautions in attack, with appropriate care. But before we leave the ethical and philosophical part of the discussion, one final question should be posed. It is this: is a conflict characterised by the use of autonomous weapon systems, whether in the form of autonomous platforms or autonomous cyber weapons, by both sides in attack and defence a war, or armed conflict at all? From a legal perspective the answer is clear and in the affirmative. The weapon systems will be armed, programmed and deployed by the armed forces of the States involved in the conflict and will be intended to cause casualties and damage to the personnel and property of the opposing party. Provided the intensity and sustainment of the violence, and the characteristics of the parties to the conflict meet the applicable criteria, such activity is capable of constituting either an international or non-international armed conflict depending on the circumstances.

So is a 'machine versus machine' clash of arms in which people remain peripheral 'war as we are prepared to recognize it'. What is war all about when machine takes on machine and people merely observe and suffer? The answer seems to be that such a conflict meeting the armed conflict threshold is what armed conflict always was, namely a contest between States or within a State in which the parties use the resources, technologies and personnel at their disposal in an effort to impose their wills on each other. It seems however that aspects of autonomous attack, and maybe of certain kinds of cyber attack, are likely to be found ethically unacceptable, and it will be for international discussions, for example of the sort being undertaken under the aegis of the Conventional Weapons Convention,[15] to determine what legal or other measures are required to address such international concerns.

[14] As national discussions associated with a proposed intervention following the alleged use of chemical weapons in Syria in 2013 demonstrate, the suggested right of humanitarian intervention is at best controversial.

[15] At the time of writing, the next meeting under the aegis of the Convention on Prohibitions or Restrictions on the Use of Certain Conventional Weapons Which may be Deemed to be Excessively Injurious or to Have Indiscriminate Effects, Geneva, 10 October 1980 (CCW), to discuss autonomous weapons took place in April 2015.

4 The Law That Article 36 Reviews Must Apply

We have noted that international law does not cite remoteness in attack as such as the basis for prohibiting a method of warfare. Any particular method of warfare must, however, be assessed against the international law rules applying to the relevant State to determine its lawfulness. In this section, therefore, we consider what those rules are, paying particular attention to the rules that seem to be of greatest relevance to RPA, cyber and autonomous weapon system operations. A core principle that binds all States provides that: "In any armed conflict, the right of the parties to the conflict to choose methods or means of warfare is not unlimited."[16] The principles and rules that we are about to discuss constitute the limits to which this core principle refers.

A cardinal customary principle in the law of armed conflict and, thus, of the law of weaponry[17] that binds all States and has its roots, inter alia, in certain nineteenth Century Declarations and in treaties adopted in 1899 and 1907[18] prohibits the employment of "weapons, projectiles and material and methods of warfare of a nature to cause superfluous injury or unnecessary suffering."[19] The practical application of this principle requires that "The legitimacy of a weapon, by reference to the superfluous injury and unnecessary suffering principle, must be determined by comparing the nature and scale of the generic military advantage to be anticipated from the weapon in the application for which it is designed to be used, with the pattern of injury and suffering associated with the normal intended use of the weapon."[20]

A second customary principle of the law as to weapons, that also therefore binds all States, prohibits weapons that are indiscriminate by nature. Contemporary understanding of the Rule is given in article 51(4)(b) and (c) of API, which provides that prohibited indiscriminate attacks include "(b) those which employ a method or means of combat which cannot be directed at a specific military objective; or (c) those which employ a method or means of combat the effects of which cannot be limited as required by th[e] Protocol; and [which], consequently, in each such case,

[16] API, article 35(1) and see Oeter (2013), pp. 121–125.

[17] ICJ Nuclear Weapons Advisory Opinion paragraphs 74–87.

[18] See for example St Petersburg Declaration, 1868, preambular paragraphs 3–6, Brussels Declaration, 1874, article 12 and Hague Regulations 1899 and 1907, article 23(e). The early workings and development of the rule are discussed at Boothby (2009), pp. 55–58; see also Solis (2011), pp. 269–272.

[19] API, article 35 (2).

[20] See Fenrick (1990), p. 500: "A weapon causes unnecessary suffering when in practice it inevitably causes injury or suffering disproportionate to its military effectiveness. In determining the military effectiveness of a weapon, one looks at the primary purpose for which it was designed." See also the discussions at Oeter (2013), pp. 125–126 and at Dinstein (2010), pp. 63–67.

are of a nature to strike military objectives and civilians or civilian objects without distinction."[21]

It is the inability to control the direction and/or effects of the weapon that renders it indiscriminate by nature and which therefore renders it unlawful.

The next two rules address the impact of warfare on the natural environment. The first is concerned with the use of the environment itself as a weapon and prohibits States party to "engage in military or any other hostile use of environmental modification techniques having widespread, long-lasting or severe effects as the means of destruction, damage or injury to any other State party."[22]

'Widespread' encompasses "an area on the scale of several hundred square kilometres"; 'long-lasting' involves "lasting for a period of months, or approximately a season"; and 'severe' requires that there be "serious or significant disruption or harm to human life, natural and economic resources or other assets."[23] The use of cyber or remotely piloted methods to attempt to modify the weather, for example by increasing or reducing rainfall in order to cause floods or drought in the territory of a State party to the treaty would constitute activity that may be capable of breaching this treaty rule.[24]

A second rule relating to the natural environment and which binds States that are party to API prohibits "methods or means of warfare which are intended, or may be expected, to cause widespread, long-term and severe damage to the natural environment."[25] Specifically, "[t]his protection includes a prohibition of the use of methods or means of warfare which are intended or may be expected to cause such damage to the natural environment and thereby to prejudice the health or survival of the population."[26] The 'widespread, long-term and severe' criteria in this rule are cumulative, so if any one of them is not present, the rule is not broken. Although none of the criteria is defined, their combined effect is that only the most serious of damage will breach the rule.[27]

---

[21]Modern examples of weapons that would breach the rule would include many of the improvised rockets fired from Gaza into Israel in recent years; see Scobbie (2012), p. 307.

[22]Convention on the Prohibition of Military or any Other Hostile Use of Environmental Modification Techniques, Geneva, 2 September 1976 (ENMOD), article I. The term 'environmental modification techniques' refers to "any technique for changing – through the deliberate manipulation of natural processes – the dynamics, composition or structure of the Earth, including its biota, lithosphere, hydrosphere and atmosphere, or of outer space."; article II. Note the additional obligation at article I(2) not to assist, encourage or induce activities that breach this provision.

[23]Conference Understanding relating to Article I, available at the ICRC treaty database at www.icrc.org.

[24]Rowe (1987), p. 117. A further Conference Understanding indicates that the possible results of such techniques may include earthquakes, tsunamis, upset in the ecological balance of a region, changes in weather and climate patterns; Conference Understanding relating to Article II, available at the ICRC treaty database at www.icrc.org.

[25]API, article 35(3).

[26]API, article 55(1).

[27]"The time or duration required (i.e. long-term) was considered by some to be measured in decades. References to twenty or thirty years were made by some representatives as being a

Those then are the weapons law rules that apply to all weapon technologies. There are then some specific law of armed conflict rules that prohibit or restrict the use of particular weapons types. These rules are numerous and complex, so in the interests of appropriate brevity, some of the more technically detailed rules must inevitably be omitted.[28]

All States are prohibited from using poison, poisoned weapons,[29] asphyxiating, poisonous or other gases, all analogous liquids, materials or devices and bacteriological methods of warfare.[30] All States are similarly prohibited from using bullets that expand or flatten easily in the human body in international armed conflicts and, in most circumstances, in non-international armed conflicts; they are also prohibited from using explosive or incendiary bullets designed solely for use against personnel, a prohibition that applies in both international and non-international armed conflicts.[31]

Fragmentation weapons the primary effect of which is to injure by fragments which in the human body escape detection by X-rays cannot be used by States that are party to the relevant treaty.[32] Two Protocols to CCW, namely Protocol II and Amended Protocol II, include some very technical and detailed prohibitions and restrictions in relation to mines,[33] booby traps[34] and other devices.[35] Protocol II,

---

minimum. Others referred to battlefield destruction in France in the First World War as being outside the scope of the prohibition . . . .It appeared to be a widely shared assumption that battlefield damage incidental to conventional warfare would not normally be proscribed by this provision. What the article is primarily directed to is thus such damage as would be likely to prejudice, over a long term, the continued survival of the civilian population or would risk causing it major health problems."; Rapporteur's Report CDDH/215/Rev.1 paragraph 27 reported in ICRC Commentary, paragraph 1454.

[28]For a more detailed consideration of these matters, see Hays Parks (2005), p. 55.

[29]Instructions for the Government of Armies of the United States in the Field (Lieber Code) 24 April 1863, article 16; Hague Regulations, article 23(a) and see Dinstein (2010), pp. 68–69.

[30]Geneva Protocol for the Prohibition of the Use in War of Asphyxiating, Poisonous or Other Gases, and of Bacteriological Methods of Warfare, 1925.

[31]Manual on the Law of Armed Conflict, UK Ministry of Defence (2004) (UK Manual), paragraphs 6.9 and 6.10 and see Dinstein (2010), pp. 69–70.

[32]CCW Protocol on Non-Detectable Fragments, Geneva, 10 October 1980 (Protocol I).

[33]"Mine means any munition placed under, on or near the ground or other surface area and designed to be detonated or exploded by the presence, proximity or contact of a person or vehicle"; Protocol on Prohibitions or Restrictions on the Use of Mines, Booby-Traps and Other Devices, Geneva, 10 October 1980 (Protocol II), article 2(1).

[34]A booby-trap is "any device or material which is designed, constructed or adapted to kill or injure and which functions unexpectedly when a person disturbs or approaches an apparently harmless object or performs an apparently safe act"; Protocol II, article 1(2) and Protocol on Prohibitions or Restrictions on the Use of Mines, Booby-Traps and Other Devices, as amended on 3 May 1996, Geneva, (Amended Protocol II), article 2(4).

[35]"Other devices means manually-emplaced munitions and devices designed to kill, injure or damage and which are actuated by remote control or automatically after a lapse of time"; Protocol II, article 1(3). Amended Protocol II altered the definition to read "Other devices means manually-emplaced munitions and devices including improvised explosive devices designed to kill, injure

which was adopted in 1980, largely reflects relevant military doctrine. The Amended Protocol makes more prescriptive provision and amends certain of the definitions. So, article 6 of Protocol II prohibits any use of a booby-trap "in the form of an apparently harmless portable object which is specifically designed and constructed to contain explosive material and to detonate when it is disturbed or approached".[36] The use of booby-traps in any way attached to or associated with any of a lengthy list of objects and facilities is also prohibited.[37]

In very general terms, Amended Protocol II prohibits mines, booby-traps or other devices designed to detonate due to the presence of commonly available mine detectors during normal use in detection operations[38]; self-deactivating mines with an anti-handling device so designed as to be able to function after the mine can no longer function[39]; and non-detectable anti-personnel mines produced after 1 January 1997.[40] It also prohibits remotely delivered mines that do not comply with certain technical self-destruction, self-deactivation and self-neutralisation requirements.[41]

Under Protocol III to CCW it is prohibited to make military objectives located within a concentration of civilians the object of attack by air-delivered incendiary weapons. Making a similarly located military objective the object of attack by non-air-delivered incendiaries is only permissible if the military objective is clearly separated from the concentration of civilians and all feasible precautions are taken to limit the incendiary effects to the military objective and to avoid or minimize loss of civilian life, injury to civilians and damage to civilian objects.[42]

The employment of "laser-weapons specifically designed, as their sole combat function or as one of their combat functions, to cause permanent blindness to unenhanced vision, that is, to the naked eye or to the eye with corrective eyesight devices" is prohibited under Protocol IV to CCW.[43]

or damage and which are actuated manually, by remote control or automatically after a lapse of time"; Amended Protocol II, article 2(5).

[36]Protocol II, article 6(1)(a).

[37]Protocol II, Article 6(1), Amended Protocol II, article 7(1).

[38]Amended Protocol II, article 3(5).

[39]Amended Protocol II, article 3(6).

[40]Amended Protocol II, article 4 and Technical Annex, paragraph 2(a).

[41]Amended Protocol II, Technical Annex, paragraphs 3(a) and (b) and article 6(3).

[42]Protocol on Prohibitions or Restrictions on the Use of Incendiary Weapons, Geneva, 10 October 1980, article 2(2) and (3) respectively. There is a lengthy and restrictive definition of incendiary weapons in article 1(1) of the Protocol. The US entered a reservation on ratification that would permit the attack of targets, such as a chemical weapons plant, which may be located in populated areas on the basis that the use of incendiaries to destroy the plant would produce fewer casualties than if the chemicals were released with explosive munitions.

[43]Protocol on Blinding Laser Weapons, Geneva, 13 October 1995 (Protocol IV), Article 1. Incidental or collateral blinding resulting from the legitimate use of laser systems is not prohibited; article 3. Permanent blindness means irreversible and uncorrectable loss of vision that is seriously disabling with no prospect of recovery. Serious disability is equivalent to visual acuity of less than 20/200 Snellen measured using both eyes; Protocol IV, article 4.

The Chemical Weapons Convention 1993[44] prohibits use, development, production, acquisition, stockpiling, retention and direct or indirect transfer of chemical weapons. Military preparations for, or operations associated with, using such weapons and assisting, encouraging or inducing anyone to do anything prohibited to a State party to the Convention are also banned.[45] Chemical weapons include: toxic chemicals and their precursors, but not where they are intended for purposes that do not breach the Convention and provided the types and quantities are consistent with such purposes; munitions and devices that are designed to cause death or other harm through the toxic properties of those toxic chemicals that would be released due to the use of such munitions and devices; equipment designed for use directly in connection with the employment of such munitions and devices.[46]

Toxic chemicals are chemicals which, through their chemical action on life processes, can cause death, temporary incapacitation or permanent harm to humans or animals. This includes all such chemicals, regardless of their origin or of their method of production, and regardless of whether they are produced in facilities, in munitions or elsewhere.[47] A 'precursor' is any chemical reactant which takes part, at any stage, in the production, by whatever method, of a toxic chemical, including any key component of a binary or multi-component chemical system.[48]

The possession of such chemicals is lawful, however, if the chemical is intended for and consistent with purposes which are not prohibited under the Convention. Permitted uses are listed and include law enforcement, including domestic riot control purposes.[49] The prohibition on the use and probably the prohibition of the possession of chemical weapons other than for the purposes listed in article II(9) are now rules of customary law binding on all States.[50] The Convention applies to both international and non-international armed conflicts and all planning and training for use of such weapons is also illegal.

Riot control agents may not be used as a method of warfare but may be used for law enforcement, including domestic riot-control, purposes. Riot control agents comprise "chemicals not listed in a Schedule to the Treaty which can produce rapidly in humans sensory irritation or disabling physical effects which disappear within a short time following termination of exposure".[51]

[44] Convention on the Prohibition of the Development, Production, Stockpiling and Use of Chemical Weapons and on Their Destruction, Paris, 13 January 1993 (Chemical Weapons Convention).

[45] Chemical Weapons Convention, Article I(1).

[46] Chemical Weapons Convention, Article II(1).

[47] Chemical Weapons Convention, Article II(2).

[48] Chemical Weapons Convention, Article II(3).

[49] Chemical Weapons Convention, article II(9). Law enforcement would appear to refer to the enforcement of the domestic law and order; Fidler (2005), pp. 540–544.

[50] See ICRC Customary Law Study (Henckaerts and Doswald-Beck 2004), Rules 74–76.

[51] Chemical Weapons Convention, 1993, Article II(7).

Article I of the Biological Weapons Convention 1972,[52] which applies to international and non-international armed conflicts, prohibits the development, production, stockpiling, acquisition or retention of microbial or other biological agents or toxins whatever their origin or method of production, of types and in quantities that have no justification for prophylactic, protective or other peaceful purposes, and weapons, equipment or means of delivery designed to use such agents or toxins for hostile purposes or in armed conflict.[53] These prohibitions are now customary law that binds all States.[54] Planning and training for use of such weapons is also illegal.

States party to the Ottawa Convention 1997[55] are prohibited to use, develop, produce, otherwise acquire, stockpile, retain or transfer to anyone, directly or indirectly, anti-personnel mines and must not assist, encourage or induce anyone to do such things.[56] An anti-personnel mine for the purposes of the Ottawa Convention is a mine designed to be exploded by the presence, proximity or contact of a person and that will incapacitate, injure or kill one or more persons. Mines designed to be detonated by the presence, proximity or contact of a vehicle as opposed to a person and that are equipped with anti-handling devices, are not considered anti-personnel mines as a result of being so equipped.[57] The prohibition applies to international and non-international armed conflicts and binds States party to the treaty.

The Convention on Cluster Munitions 2008 prohibits the use, development, production, acquisition, stockpiling, retention or direct or indirect transfer of cluster munitions or assistance, encouragement or inducement of such activities.[58] The Convention, the main focus of which is explosive cluster munitions, includes a complex definition that specifically excludes certain technologies.[59]

These, then, in very brief outline form are the core provisions of weapons law. As the language of the weapons review obligations in article 36 of API makes clear, it is existing law that provides the yardstick against which the lawfulness of any

[52]Convention on the Prohibition of the Development, Production and Stockpiling of Bacteriological (Biological) and Toxin Weapons and on their Destruction 1972, was opened for signature on 10 April 1972 (Biological Weapons Convention).

[53]Biological Weapons Convention, Article I. The Convention does not, in terms, prohibit use of biological or bacteriological weapons or materials, but during the Fourth Review Conference in 1996 it was agreed among the states party that Article 1 has the effect of prohibiting the use of such weapons; UK Manual page 104 at note 8.

[54]ICRC Customary Humanitarian Law Study (Henckaerts and Doswald-Beck 2004), p. 256.

[55]Convention on the Prohibition of the Use, Stockpiling, Production and Transfer of Anti-Personnel Mines and on their Destruction, Oslo, adopted on 18 September 1997 (Ottawa Convention).

[56]Ottawa Convention, article 1.

[57]Ottawa Convention, article 2(1).

[58]Convention on Cluster Munitions, Dublin, 30 May 2008 (Cluster Munitions Convention), article 1(1). Note, however, the special provisions in Article 21 that address interoperability between, respectively, States that are and are not party to the Convention.

[59]Cluster Munitions Convention, Article 2(2).

new weapon, means or method of warfare must be judged. How, then, do these rules translate into legal criteria to be applied in a weapon review?

5 The Legal Criteria to be Applied in a Weapon Review

The article 36 obligation to review the lawfulness of a weapon, means or method first arises with its study. The study must be of a weapon, means or method, not, for example, of a technology that might at some point in the future be capable of development into a weapon, means or method. It is a matter of interpretation when the study of a novel technology becomes the study of a weapon. Perhaps it is when weaponisation is first being evaluated.[60]

Article 36 also mentions 'development', which seems to include application of materials, equipment, and other elements to form a weapon and the improvement, refinement and probably the testing of prototype weapons with a view to achieving optimal performance.[61] 'Acquisition' may be from commercial undertakings and/or other states, for consideration or by way of gift, and 'adoption' involves a state or its armed forces deciding to use a particular weapon or method of warfare in military operations. Weapons must be reviewed for legal compliance irrespective of whether another state has previously conducted its own weapons review.

When undertaking a weapon review, a State assesses the general circumstances in which it is intended to use the weapon and determines whether the existing rules of law applicable to that State prohibit or restrict those general intended circumstances of use. If they do, this will be made clear in the weapon review. International law does not prescribe any particular form that a weapon review must take, nor is there a set procedure to be adopted in preparing it. Advice to an appropriate commander may, depending on the circumstances, be sufficient.[62]

The criteria that a State should apply when conducting a weapon review are:

1. Is the weapon, means or method of warfare of a nature to cause superfluous injury or unnecessary suffering? The practical application of the test has been authoritatively explained as follows:

   "The prohibition of unnecessary suffering constitutes acknowledgment that necessary suffering to combatants is lawful and may include severe injury or loss of life. There is no agreed international definition for unnecessary suffering. A weapon or munition would be deemed to cause unnecessary suffering only if it inevitably or in its normal use has a particular effect and the injury caused is considered by governments as disproportionate to the military necessity for it, that is, the military advantage to be gained from its use. This balancing test

[60] Boothby (2009), p. 345.

[61] Consider Daoust et al. (2002), p. 348.

[62] Tallinn Manual, commentary accompanying Rule 48, paragraph 3.

cannot be conducted in isolation. A weapon's or munition's effects must be weighed in light of comparable, lawful weapons or munitions in use on the modern battlefield. A weapon is not unlawful merely because it may cause severe suffering or injury. The appropriate determination is whether a weapon's or munition's employment for its normal or expected use would be prohibited under some or all circumstances. The correct criterion is whether the employment of a weapon for its normal or expected use inevitably would cause injury or suffering manifestly disproportionate to its military effectiveness."[63]

So what are 'comparable' weapons where, for instance, a cyber tool is developed to fulfil a military requirement normally achieved, say, by using a conventional bomb. The proper comparator in such a case must be the means that would, absent the new technology, otherwise be employed. This will inevitably involve comparing unlike damaging and injuring mechanisms but seems to be the only sensible way to apply the customary rule.

2. Is the weapon, means or method of warfare by its nature indiscriminate? The issue here is whether the weapon, means or method will inevitably act in an indiscriminate way either because it cannot be directed at a specific military objective or because its damaging effects cannot be limited, as required, to the military objective.
3. For a state party to API, is the weapon, means or method of warfare intended, or may it be expected, to cause widespread, long-term and severe damage to the natural environment? It is for individual States to decide whether voluntarily to impose on themselves more stringent limits than this API criterion would imply.[64]
4. For a State party to the Environmental Modification Convention 1976, does the weapon, means or method of warfare constitute military or any other hostile use of an environmental modification technique that may be expected to have widespread, long-lasting or severe effects as the means of destruction, damage or injury to any other State that is party to the treaty?
5. A weapon review should then apply the rules relating to specific weapons and weapons technologies that were briefly summarized in Sect. 4 and that are relevant to the weapon, means or method of warfare that is being reviewed. It should be appreciated that the applicable rules are more detailed than the space allocated to Sect. 4 permits, and the relevant treaties and other authorities should therefore be consulted. If the planned procurement contemplates the use of a weapon in circumstances that would or may conflict with weapons law restrictions, the weapon reviewer will need to draw attention to the relevant legal restrictions in the review.

[63] Reproduced in Hays Parks (2006), at note 25.

[64] Moreover, it is conceivable that environmental considerations will over time impinge more on military activities; see for example UK Ministry of Defence, DCDC, Future Maritime Operational Concept 2007, 13 November 2007, paragraph 122.

6 How Do Specific Remote Attack Technologies Match Up to These Criteria?

Having established the legal criteria that must be applied during a weapon review, and having shown that it is the current rules of international law that States are obliged to apply, the question we should now address is how these criteria apply to the new technologies that were mentioned earlier in this paper. We concluded that remoteness in attack *per se* does not raise international law issues. We will therefore take the individual technologies in turn in order to determine in what respects they may be challenged by the applicable legal rules. We will apply the two customary principles first, followed by the two environmental protection rules. Thereafter we will assess whether there are any rules that have specific relevance to the particular technology.

Some of the technologies that we are discussing will, however, require the consideration of additional legal rules as part of the legal review process. This is because autonomous weapon systems and some cyber capabilities will not involve a human being in making individual attack decisions. That novel feature makes it necessary for a person conducting a legal review of such technologies to consider whether the technology in question, and the intended method of using it, will enable the rules of targeting law to be complied with.

Having addressed that aspect of each of the three novel weapon technologies, we will seek to draw some general conclusions.

6.1 Remotely Piloted Aircraft

The superfluous injury and unnecessary suffering rule is, somewhat self-evidently, most concerned with the injury, and suffering caused to personnel belonging to the adverse party to the conflict. The nature of the platform that is used to deliver the relevant munition, the way in which that platform is guided, and the circumstances in which a decision is made to fire the munition carried by the RPA are matters that are certainly of legal interest, but which do not generally raise issues under the superfluous injury and unnecessary suffering principle. In a legal review of an RPA, therefore, it will not generally be necessary to consider the principle in any detail, other than perhaps to explain the reason for its irrelevance.

In determining whether an RPA weapon system is indiscriminate by nature, that is whether it is incapable of being directed at a military objective or is unable reasonably to limit its injurious or damaging effects to the intended target, due account will have to be taken of a number of matters. These include the various on- and off-board sensors that provide data to support the attack decisions made by the RPA operator, the computerized links that convey that data to the RPA operator, the manner in which that data is processed and presented to the operator, the up-links that convey the RPA operator's instructions to the platform, the tasking of and support provided to the RPA operator as well as the wider way in which such missions are planned and executed. The robustness of such computerized up- and down-links will also be a relevant factor. Generally speaking, it would seem to be highly unlikely that an RPA weapon system will be assessed as indiscriminate by nature. This is simply because it will usually be capable of discriminate use. However, if tests or other data indicate that there are circumstances in which the ability to use such a system discriminately will foreseeably be eroded, appropriate comment in that regard should be included in the weapon review. So, for example, if there are fears that the up- or down-links to such a platform may be susceptible to interference by the enemy or others, it would be prudent to consider whether the technology should be required to disclose to the operator when such interference is occurring and should, to the extent possible, protect the weapon system against such interference.

When addressing the API environmental protection rule, similar considerations apply as were considered in this section in relation to superfluous injury and unnecessary suffering. The nature of the platform that is used to deliver the relevant munition, the way in which that platform is guided, and the circumstances in which a decision is made by the operator to fire the munition carried by the RPA will not generally raise issues under articles 35(3) and 55 of API and a legal review of an RPA can usually, therefore, simply state that fact.

There are no rules of the law of armed conflict that specifically apply to RPAs or to equivalent remotely controlled platforms on land or in the maritime environment. Moreover, because a person is in the controlling loop of such weapon systems, determining where the platform will go and making its attack decisions, there will not be a weapon review requirement to consider the ability of the weapon system to comply with the rules of targeting law. This is because the RPA operator will, like the pilot of a manned aircraft, be responsible for ensuring that attack decisions comply with the distinction and discrimination principles, with the proportionality rule, with special protection rules and with the rules as to precautions in attack.[65]

It therefore seems likely that the main focus of an RPA weapon review will be on the indiscriminate weapons principle and that this is a principle that such weapon systems will generally have little difficulty in satisfying.

6.2 Autonomous Weapon Systems

We shall now consider the issues that will arise in the weapon review of an autonomous or highly automated weapon system. The superfluous injury and unnecessary suffering principle is generally going to be of greatest relevance to the damaging, wounding or injuring effect of the munition or projectile that the weapon system releases or fires, as opposed to the technology that identifies the target that is

[65]Consider e.g. articles 48–67 of API.

to be attacked. Accordingly, in a legal review of the autonomous weapon's guidance system, as opposed to the munition or projectile it guides, the superfluous injury and unnecessary suffering criterion will usually be of little relevance and it is not necessary for us to consider it further in the present sub-section.[66]

Deciding whether the autonomous weapon system complies with the indiscriminate weapons rule will necessarily involve assessing the actual performance of its autonomous target recognition technology. Testing will be required in order to inform an assessment of the performance of those features and, if available, battlefield data will also need to be reviewed. Clearly, those who designed the weapon system will have sought to ensure that it reliably recognizes objects or persons that are lawful objects of attack.

Reliable autonomous recognition of persons under combat conditions as lawful targets of attack is unlikely to be feasible in the short term, possibly not even in the medium term. Autonomous recognition of items of military hardware such as tanks, artillery pieces, armored personnel carriers etc. may, however, be realistically possible using currently available technology.

Clearly, the technology will be designed in order to try to achieve the required military purpose, namely to identify military objects that are the desired objects of attack. Accordingly it will be seeking to differentiate between objects that can lawfully be attacked and those that are protected from attack by the law. Data as to the performance of the autonomous target recognition system will establish whether in fact the technology is capable of achieving this differentiation to a sufficient degree of reliability to comply with the discrimination rule in article 51(4) of API. It should, however, be recalled that a weapon will only be indiscriminate by nature if it cannot be directed at a specific military objective or if its effects cannot be limited as required by international law and if the result in either case is that the nature of the weapon is to strike military objectives and civilians or civilian objects without distinction.

It would seem to be sensible to consider the problem by reference to three possible examples of such technology. Imagine first an autonomous attack system that is designed to recognize particular features or characteristics of, say, an artillery piece, tank or other item of military equipment and that has been shown to perform that recognition task satisfactorily in tests. Imagine that these tests realistically represent the general circumstances in which it is intended to use the weapon system. Such a technology would seem, for the purposes of the indiscriminate weapons rule, to be capable of being directed at a military objective. It will be the characteristics of the munition or projectile that is being fired, and perhaps its influence on particular kinds of target, that will determine whether the effects of the weapon are capable of being limited as required by article 51(4)(c).

Consider, secondly, an autonomous attack system designed to recognize as a target, and to attack, any vehicle above a particular gross weight, or with a metal

---

[66] See Anderson and Waxman (2013), pp. 10–11; and Schmitt and Thurnher (2013), pp. 244–245 as to the application of the superfluous injury principle to autonomous weapon systems.

content above a specified minimum, that appears in its area of search. If civilian heavy goods and public service passenger vehicles in the area being searched by the weapon system would satisfy the pre-set algorithms and thus be recognized by the system as targets, careful evaluation of the autonomous weapon system would be required. Such an autonomous weapon system may be capable of discriminating use in particular circumstances, for example if its attack options are suitably constrained at the mission planning stage. This might for example be achieved by limiting its permitted area of search to locations where civilian vehicles satisfying its target algorithms are known not to be. However, the weapon review must draw attention to the restricted circumstances in which its employment would be legitimate[67] and should set out the actions that will be required in order to seek to ensure that when the weapon system is used the discrimination principle will be complied with.

For essentially similar reasons to those discussed in the previous sub-section in relation to RPAs, the treaty-based environmental protection rules are unlikely to be relevant to the autonomous target identification aspect of a novel weapon system.[68] Moreover, the treaty and customary rules of the law of armed conflict include no provision that refers explicitly to autonomous attack technology.[69]

We noted earlier in this section that a weapon review of a weapon system employing autonomous attack technology should not limit itself to a consideration of the usual weapons law criteria summarised in Sect. 5, but should also consider whether the technology, and its intended manner and circumstances of use, enable the rules of targeting law to be complied with. In the following paragraphs, some aspects of targeting law will be considered in order to illustrate how a weapon review should, in the author's view, address this aspect of such a weapon system.

Importantly, if a weapon system employs autonomous attack technology in such a way that compliance with specific targeting law rules is rendered impracticable, the legal review should refer to this and should explain the considerable limiting effect this is likely to have on the circumstances in which the weapon system can lawfully be used. The precautions required of an attacker in article 57 of API probably illustrate most clearly the challenges that are likely to be posed for certain highly automated and autonomous attack technologies. They will not necessarily be an exhaustive list of the aspects of targeting law that the weapon reviewer of an

[67]In this regard, recall the reference to use in some or all circumstances in article 36 of API which would extend to the corresponding implied rule. As to the application of the indiscriminate weapons rule to autonomous attack technology, see Schmitt and Thurnher (2013), pp. 245–250 and as to the weapons review of such systems, see pp. 271–276.

[68]If, however, such an autonomous weapon system were to be designed to undertake attacks which may be expected to occasion the environmental damage referred to in the treaty rules, for example attacks on nuclear electricity generating stations or on certain kinds of chemical plant, the rules in articles 35(3) and 55 should be addressed in the weapon review.

[69]S.S. 'Lotus' (Fr. v. Turk.), 1927 P.C.I.J. (ser. A) No. 10, (Sept. 7) p. 18 established the principle that restrictions upon the independence of States cannot be presumed. Applying this principle, the absence of specific reference to such technologies leads to the conclusion that they are not the subject of a prohibition.

autonomous weapon system will need to consider, but a discussion of their potential application will demonstrate the sort of factors that the reviewer will need to bear in mind.

The weapon reviewer will recall, and will wish to draw attention to, the general obligation to take constant care to spare the civilian population, civilians and civilian objects, noting that the ensuing obligations must be construed in that context.[70]

Reference should then be made in the weapon review to the detailed precautions that must be undertaken by those who "plan or decide upon an attack". These obligations apply equally to planners of automated or autonomous missions and therefore to all those involved in mission preparation, in programming software, in review of available information, in prescribing the areas of search and the times when such searches are to take place, in setting the target identification criteria for the weapon control software and so on. As this list illustrates, people will continue to be of central importance to the successful completion of missions and it may be prudent to draw attention to this aspect, and to the constant care they are required to exercise, in the weapon review document.[71]

The mere fact that it is the machine that decides to engage a particular target does not mean that the precautionary obligations of decision makers in article 57(2)(a)(i) to (iii) cease to apply. Rather, the weapon reviewer must be satisfied that the autonomous weapon system and its method of employment enable the relevant precautions to be undertaken, whether through action taken by the equipment itself or because of what personnel operating the weapon systems, supervising them or planning the sortie are enabled to do in advance of, or during, the sortie.

Article 57(2)(a)(i) and (ii) requires everything 'feasible' to be done to fulfil its requirements. So if one of those obligations can only be fulfilled using a manned mission, the manned mission should be employed. This may become an issue if an autonomous or highly automated weapon system cannot be programmed to recognize when its employment would preclude the taking of sub-paragraph (i) or (ii) precautions that could be taken if some other weapon system, such as a manned one, were to be employed. The inability of the autonomous or automated system to fulfill the obligations does not render them non-feasible because it would be feasible for a human decision-maker to fulfil them. Precautions that can be fulfilled using the weapon systems more traditionally used to undertake the relevant kind of attack are indeed feasible and should therefore be taken. An autonomous or automated weapon system that is not capable of making that determination (that one of the 57(2)(a)(i) or (ii) obligations can only be fulfilled using a non-autonomous/automated system) and of acting accordingly fails to meet the precautions obligations and should either not be procured or at least should not be employed in the relevant circumstances.

---

[70] API, article 57(1).

[71] Multinational Capability Development Campaign (2013–2014), Role of Autonomous Systems in Gaining Operational Access, Policy Guidance (MCDC Policy Guidance)—Autonomy in Defence Systems, 29 October 2014 (MCDC Policy Guidance), 23-5 lists some human factors worthy of consideration.

The specific obligations under article 57(2)(a)(i) to do everything feasible to verify that the object of the attack is a military objective and that it is not entitled to special protection[72] are vital to compliance with the principle of distinction. If, say, an algorithm-based technology is found in tests satisfactorily to differentiate between the objects it is programmed to identify and those it is not, i.e. between military objects and civilian objects, it will probably satisfy this precautionary obligation.[73] If automated or autonomous target recognition technology is developed with a view to identifying, and attacking, human targets, there will be considerable challenges under this first element of article 57(2)(a)(i). Can it, in particular, be shown that the target identification software or mechanism satisfactorily differentiates between, on the one hand, lawful targets, i.e. able-bodied combatants[74] or able-bodied civilians directly participating in the hostilities, and, on the other hand, persons whom the law protects, i.e. combatants who are hors de combat, non-combatants and civilians who do not directly participate.[75] Technical developments in this field may be difficult to predict but the legal issue is clear. Can the autonomous or automated weapon system, when employed in the intended manner in an anti-personnel role, distinguish between combatants and civilians, between combatants and non-combatants, between able-bodied combatants and those who are hors de combat, between directly participating civilians and those not participating and between directly participating civilians who are able-bodied and those who are hors de combat?

Article 57(2)(a)(i), however, imposes additional precautionary obligations. It requires that attackers do everything feasible to verify that it is not prohibited by the Protocol to attack the intended targets. Examples of attacks that would breach the Protocol include attacks that would breach articles 51(4) (discrimination principle), 51(5)(a) (separate and distinct military objectives treated as one), 51(5)(b) (proportionality), 53 (cultural objects), 54 (objects indispensable to the survival of the civilian population), 35(3) and 55 (protection of the natural environment), 56 (works and installations containing dangerous forces and military objectives in their vicinity), 41 (safeguarding of persons hors de combat), 12 and 15 (protection of medical units and personnel) and 21 to 28 (protection of medical transports).

---

[72]See AMW Manual, Rule 32(a).

[73]The reference here to 'military objects' is intentional; the technology is likely to be configured so as to distinguish, for example, between an artillery piece or a tank on the one hand and a civilian vehicle of comparable size on the other; see Lewis et al. (2009), p. 10 and Wagner (2011), EAP7.

[74]Note, for example, that if, as seems possible, nanotechnology enables the development of military firearms and equipment with reduced metal content, an autonomous or automated weapon system that would differentiate between combatants and civilians only on the basis of their metallic signature may prove problematic.

[75]The Adversary Behavior Acquisition, Collection, Understanding and Summarization facility integrates data from a number of sources and would apply a human behaviour modeling and simulation engine to generate intent-based threat assessments associated with individuals and groups; Schachtman, Army Tracking Plan: Drones that Never Forget a Face, 28 September 2011 available at www.wired.com/dangerroom/2011/09/drones-never-forget-a-face/.

Where these rules simply prohibit attacks directed at the relevant object or person, the relevant question will be whether the automated or autonomous weapon system, when used in its designed or intended manner, will detect that a prospective target comes within one of these protected categories and whether it will accordingly refrain from attacking it. It is not at the time of writing clear whether software can be developed that distinguishes, for example, between an able-bodied combatant and one who comes within article 41 as being hors de combat.[76]

We should now move on to consider the evaluative precautions involved, for example, in article 51(5)(a), in the proportionality assessment referred to in articles 51(5)(b) and 57, and in the tests in article 57(2)(a)(ii) and 57(3) of API. Article 57(2)(a)(ii) requires that all feasible precautions be taken "in the choice of means and methods of attack with a view to avoiding, and in any event to minimizing, incidental loss of civilian life, injury to civilians and damage to civilian objects". Discharging this obligation when employing autonomous or highly automated weapon systems will require that a decision is made, for example, as to whether an attack should be undertaken using a manned, operator-controlled, automated or autonomous platform with a view to minimizing incidental civilian injury and damage. The weapon reviewer will need to be satisfied that the available technology makes or facilitates the making of assessments as to the incidental civilian injury and damage to be expected from the alternative possible methods of undertaking the attack, as to which method minimizes these and that the software will contribute positively to ensuring that the appropriate method is chosen.

Article 57(2)(a)(iii) requires that those who plan or decide upon attacks do not decide to "launch any attack which may be expected to cause incidental loss of civilian life, injury to civilians, damage to civilian objects, or a combination thereof, which would be excessive in relation to the concrete and direct military advantage anticipated." Article 57(2)(b) requires that an attack be cancelled or suspended if it becomes apparent either that the objective is not a military one or that the attack will breach the proportionality rule that has just been cited in the present paragraph. These rules pose vitally important challenges for autonomous and highly automated weapon systems. Can, for example, the software of an autonomous or highly automated weapon system evaluate the military advantage to be anticipated from the destruction of a specific military object in the particular circumstances of the prospective attack? Can the autonomous or highly automated weapon system determine what incidental injury and damage to civilians and civilian objects are to be expected and can the software associated with such a weapon system compare the anticipated military advantage with the expected incidental injury and damage with a view to reaching a conclusion as to the proportionality of the prospective attack? So far as is known currently available software is not capable of undertaking such evaluative decisions, and it seems likely that for the foreseeable future autonomous

[76] As to relevant engineering challenges, see Backstrom and Henderson (2012), pp. 510–513 and for a summary of the legal issues associated with autonomous weapon systems, see MCDC Policy Guidance, pp. 14–18.

and highly automated weapon systems that are to be employed in offensive attack roles will not be capable of complying with these evaluative precautionary rules.

A distinction should at this stage be drawn between the use of autonomous or highly automated weapon systems in what may be described as 'offensive' and 'defensive' attack roles. It is well appreciated that the definition of attack in article 49(1) of API draws no distinction between offensive and defensive uses of violence. However, if an autonomous or highly automated weapon system proactively seeks out and attacks targets it can for these purposes be regarded as fulfilling an 'offensive' role, and ensuring that it complies with the precautionary rules we have discussed will clearly be challenging. Employment of such weapon technologies in an essentially defensive posture, for example to protect land-based objects and/or persons by intercepting inbound rocket systems or to protect a naval vessel by intercepting inbound torpedoes may, in the author's view, be seen differently. If such systems can adequately distinguish between the weapons they are designed to intercept and civilian objects (such as airliners or civilian vessels) and if their employment does not raise proportionality rule concerns, they may well be capable of use in accordance with the precautionary rules we have been discussing.

Moreover, it should be borne in mind that the lawfulness of new weapons, means and methods of warfare should always be assessed by reference to the designed and/or intended circumstances of their use. Imagine, for example, an autonomous or highly automated weapon system developed for use in remote and sparsely populated areas such as deserts or isolated locations at sea. Let us assume that the geographical areas where, and times during which, the autonomous or highly automated weapon system is permitted to operate are restricted such that the military value in attacking the targets the system is designed to engage will always comfortably exceed the incidental injury to civilians or damage to civilian objects that can be expected. In such circumstances, the proportionality evaluation that is made in advance of the sortie by human decision-makers may enable such an autonomous or highly automated sortie to be lawfully undertaken.

Before concluding this part of the discussion, it should be noted that article 57(2)(c) and 57(3) requires additional precautions to be taken. For reasons of available space it has not been possible to discuss these in the present paper. A weapon reviewer of an autonomous or highly automated weapon system will, however, have to consider the ability of the weapon system, in its intended circumstances of use, to comply with these rules and with the other rules of targeting law listed earlier in this sub-section.

So, in summary, if human operators are enabled to make the necessary evaluations and thus take the required precautions in a way that appropriately takes account of the circumstances of the attacks the autonomous/automated weapon system is going to undertake, this is likely to overcome the potential legal difficulties presented by article 57.[77] Absent such human involvement, the weapon reviewer

[77]Performance of the system cannot be determined in advance with certainty. The requirement is simply that these matters be addressed and that such restrictions be developed as are necessary to

will need to be satisfied that the weapon system itself can comply with the demanding requirements we have been discussing.

For the foreseeable future, therefore, those undertaking weapon reviews of weapon systems with autonomous or automated attack capabilities will wish to ensure that the way in which it is planned to use such systems is such that required precautions can be taken by personnel,[78] either before the deployment or launch of the platform or, more likely, by enabling a weapon system operator to monitor the decisions that the weapon system is making and to intervene and cancel those decisions in appropriate circumstances.

Such reviews should, however, recognize that the sophistication of these technologies will improve with time and research and that weapons technologies will usually be employed in concert with other systems and capabilities. The reviewer will need to determine what these additional systems and capabilities are and to what extent they contribute to the ability to use the system under review in compliance with the legal principles and rules that we are discussing. Nevertheless, current technology is such that a person must be in a position to cancel autonomous and some automated attack operations if the need should arise. That person will need to remain sufficiently engaged, suitably located and appropriately tasked in order to be able to observe what is taking place and, if necessary, to over-ride attack decisions reached by the weapon system. The foreseeable workloads of such personnel must be limited so as to ensure that proper decisions are made, taking all available information into account.[79]

6.3 *Cyber Weapons*

To those who believe that the idea of a weapon necessarily presupposes the immediate and direct application of kinetic force, the notion that a cyber capability can be a weapon will seem at first glance unacceptable. The two essential ingredients of a weapon, however, seem to be that the relevant equipment is an offensive capability and that it is applied, intended or designed to be applied to a target to cause injury or damage. There is no requirement as to the direct application of kinetic force; indeed, the use of biological agents or toxins, of chemical agents, of asphyxiating gases and so on in war, none of which presupposes a direct application

ensure that the circumstances in which the automated or autonomous system is used are such that its use will likely comply with the discrimination and precautions rules.

[78]Because, David Akerson notes, robotics do not, and will not in the foreseeable future, possess human judgment and discretion; Akerson (2013), pp. 69–70. Note that the University of Birmingham Policy Commission Report, The Security Impact of Drones: Challenges and Opportunities for the UK, October 2014 available at http://www.birmingham.ac.uk/Documents/research/policycommission/remote-warfare/final-report-october-2014.pdf, 7, comes to similar conclusions.

[79]This may have implications, as Peter Singer suggests, for plans that a single operator should control multiple platforms; Singer (2011), p. 342.

of kinetic force, has nevertheless long been recognized as the use of a weapon. Mike Schmitt concluded that it is the violent consequences of a cyber operation that are critical to the notion of a 'cyber attack'.[80] If one accepts that analysis, it follows logically that if a cyber capability is used, designed or intended to cause violent consequences in the form of injury or damage during an armed conflict, that cyber capability is a cyber weapon.[81]

Weapon reviews under article 36 of API, or indeed under the corresponding customary rule, are normally conducted by an individual, team or committee appointed specifically for the purpose. The determination as to the legality or otherwise of the weapon and of its generic proposed uses is normally made before the weapon is fielded, with the result that commanders know in advance of its employment that the weapon issued to them has already been judged compliant with their State's legal obligations.

A cyber weapon is, by contrast, likely to have been developed with a specific cyber attack on a known objective in mind. It will be designed to reach and attack the appropriate network or node, for example, and to have the desired effect on it. So weapons law advice as to the legitimacy of the proposed cyber weapon is likely to be required by the operational commander from his or her legal adviser at the same time as the latter gives targeting law advice as to the lawfulness of the specific, proposed attack. It would therefore seem that legal advisers to operational commanders who are likely to employ such cyber attack capabilities must be sufficiently conversant with the relevant weapons law rules to be able to advise whether the cyber weapon, as well as its intended application, are lawful.[82]

All of the elements that constitute the cyber weapon system, including the malware, software and hardware, must to the extent that they are relevant be addressed in the weapon review. The hardware equipment will not usually be of greatest importance as the legality of the cyber weapon will usually be determined by the characteristics and effect(s) of, e.g., the malicious logic.[83]

---

[80]Schmitt (2011), pp. 93–94.

[81]Note the description of 'cyber weapons' in the Tallinn Manual as "cyber means of warfare that are by design, use, or intended use capable of causing either (i) injury to, or death of, persons; or (ii) damage to, or destruction of objects, that is, causing the consequences required for qualification of a cyber operation as an attack"; Tallinn Manual, commentary accompanying Rule 41, paragraph 2.

[82]The legal advisers referred to here are those deployed to operational commanders at appropriate levels of command in accordance with article 82 of API. For the UK doctrine on the provision of legal support to deployed operations, see Legal Support to Joint Operations, Ministry of Defence Development, Concepts and Doctrine Centre, JDP 3-46 dated August 2010 (UK Ministry of Defence 2010).

[83]Malicious logic is, so far as relevant, defined in the Glossary to the Tallinn Manual as "[i]nstructions and data that may be stored in software, firmware, or hardware that is designed or intended adversely to affect the performance of a computer system. The term 'logic' refers to any set of instructions, be they in hardware, firmware, or software, executed by a computing device. Examples of malicious logic include Trojan horses, rootkits, computer viruses, and computer worms."

The nature of the weapons law advice will also tend to be different when it is a cyber weapon that is being considered. A weapon review of a non-cyber weapon will generally be concerned with that weapon's generic military utility in its normal, designed manner and circumstances of use. The targeting law decision is by contrast usually concerned, for example, with the military utility to be obtained from the use of the non-cyber weapon on a particular occasion against a particular target in a particular operational and situational context.

The fact that cyber weapons will often be designed with a particular attack on a specific military objective in mind affects both the arrangements for giving legal advice and the content of that advice. Generically based weapons law advice will probably be insufficient. Rather, the weapons law evaluation of a cyber tool designed for a particular attack on a specified target must take account of the circumstances peculiar to the planned attack in determining whether the weapons law criteria, outlined earlier in this paper, are met.

Moreover, if the cyber weapon involves attack decisions that are not made by a person, i.e. if it involves autonomous attack decision-making, then the weapon review should include a discussion of whether the weapon system is capable of use in accordance with targeting law. The factors to be considered and the reasons why they should be considered would be essentially the same as those discussed in the previous sub-section in the context of autonomous weapons.

A cyber weapon is likely to have numerous orders or levels of effect and these must all be considered when the combined weapons law and targeting law advice is being prepared on such a weapon. The first level will be the effect the cyber weapon has on the data in the target node, network or computer or its effect on the ability of the affected router to move legitimate data. The impact that such data alteration has on the performance of the targeted computer system is the second level of effect. That altered performance is liable to affect the facility that the targeted computer serves; this is the third level of effect. The injury, damage or destruction suffered by the persons or objects that are the customers of that affected facility constitute the fourth level of effect. All such levels of effect that are attributable to the use of the cyber weapon must be considered when reviewing the cyber weapon and its planned use.[84] So, if at any of those levels of effect, the cyber weapon will inevitably cause pain, wounds, other injuries or suffering that are superfluous or unnecessary in the sense discussed in Sect. 4 of this paper, the superfluous injury and unnecessary suffering test will have been broken.

A cyber weapon the effects of which can be limited to a computer node, network or system that is a military objective will not breach the indiscriminate weapons

[84]The effects to be considered will not necessarily be limited to those reflected in the suggested levels of effect. If, for example, a cyber weapon will inevitably interrupt the normal service from computers other than the targeted computer, for example computers in a botnet, and if it will thus cause injury or damage, that factor will need to be considered in determining e.g. whether the cyber weapon is indiscriminate by nature.

rule. Consider the Stuxnet attack against Iran[85] in which systems that were not the apparent object of the cyber attack were allegedly infected by the malware. However, the damaging effect was reportedly limited to the systems that were the apparent object of the attack.[86] This would lead one to conclude that the Stuxnet cyber weapon seems to have complied with the discrimination rule because mere non-damaging, non-injurious infection of other systems is insufficient to amount to a breach of the rule.[87] Cordula Droege makes the good point that "[w]hile the spread of the virus far beyond the intentions of its creators might not have caused any damage, it shows how difficult it is to control that spread."[88] So this is a factor to be considered when a weapons review of a similar cyber capability is prepared.

An alternative example might be a cyber weapon that is designed to deposit malware, i.e. malicious logic, onto a targeted website that serves both military and civilian users. If, say, all computers used to connect to that website become infected by the malware and suffer damage, meaning loss of functionality requiring replacement of physical components,[89] such a cyber weapon is likely to breach the rule. Similarly, if at any of the previously discussed levels of effect, the nature of the cyber weapon is to cause indiscriminate damage, it would also breach the rule.

The fact that computers are used to undertake an attack is per se unlikely necessarily to involve environmental impact, but a particular kind of cyber weapon might do so. Take as an example the use of cyber methods to target a nuclear electricity generating station with the purpose of causing the release of nuclear contaminants as a result of an explosion of its radioactive core. Such an attack is likely to result in environmental damage that may well reach the exacting threshold prescribed by article 35(3) of API. So it is the nature of the target and of the attack on it that are likely to have environmental consequences rather than the fact that cyber methods were employed.

There are no explicit rules of the law of armed conflict that specifically permit, prohibit or restrict the lawful circumstances of use of cyber weapons. Some of the weapon-specific rules summarized earlier in this Chapter may however, need to be considered when particular kinds of cyber weapon are subjected to review. Take as an example a cyber weapon, which is designed to insert a kill switch into the computer system controlling distribution of electrical power. Imagine that activation of the kill switch, which is designed to occur when the targeted computer is switched on, stops electricity distribution and causes power blackouts. The cyber attacker is aware that there are no back-up generators and certain consumers are expected to

---

[85]Fildes, Stuxnet worm attacked high value Iranian assets, BBC News, 23 September 2010 at http://www.bbc.co.uk/news/technology-11388018 and Broad et al., Israeli Test on Worm Called Crucial in Iran Nuclear Delay, 15 January 2011, available at http://www.cfr.org/iran/nyt-israeli-test-worm-called-crucial-iran-nuclear-delay/p23850.

[86]Richmond (2012), pp. 860–861.

[87]Tallinn Manual, Commentary accompanying Rule 43, paragraph 5.

[88]Droege (2012), p. 571.

[89]Tallinn Manual, Commentary accompanying Rule 30, paragraphs 6 and 10.

suffer injury or death as a result of the power cuts to, say, hospital intensive care units. It would be for the weapon reviewer of such a capability to determine whether such a device is a cyber booby trap[90] with the consequence that use of such a cyber weapon by a State that is party to Protocol II and Amended Protocol II to CCW must comply with articles 3, 4, 6, 7 and 8 of Protocol II and articles 3, 7, 9 and 12 of Amended Protocol II.[91]

As we saw earlier in the paper, the same treaties define and regulate 'other devices'. Let us imagine the following scenario. A thumb drive is used to insert a cyber weapon, including a kill switch, into a targeted computer system; the cyber attacker retains control over the kill switch, activates it and thereby causes an electricity distribution system to stop; the result of this is that power blackouts occur causing death and/or injury. Whether such a cyber weapon constitutes an 'other device' for the purposes of Protocol II and/or Amended Protocol II will likely depend on whether insertion by means of a thumb drive is considered to be the same thing as 'manual emplacement' in the treaty definitions. Arguably it is not, because in the example what is being emplaced by hand is the thumb drive not the malware, and it would seem to be the malware that constitutes the 'device' referred to in the treaty definition, but this is a fine distinction and it will be for a State, and its weapon reviewers, to interpret these definitions.

Additional aspects of weapons law might become relevant if cyber methods are employed to take control of an enemy platform, such as an RPA, with a view to using the weapon that the RPA is either carrying or controlling against the enemy. Let us consider a situation in which State A has taken control of State B's RPA which is carrying an airfield denial weapon that includes anti-personnel landmines as part of the composite weapon. State B is not a party to the Ottawa Convention whereas State A is a party. Having obtained control of the RPA, if State A were to use the airfield denial weapon against one of State B's airfields, this would breach State A's obligations under the Ottawa Convention.[92]

So a cyber weapon may render ad hoc weapons law rules relevant if it is used to take control of weapons to which specific rules, such as those summarised in Sect. 4 of this paper, apply.

Similarly, if it becomes possible by cyber means to take control of the enemy's weapon and then redesign the way in which it operates, this is likely also to have important legal implications. This is because design purpose is a central element

[90]Tallinn Manual, Rule 44 and accompanying Commentary.

[91]While a cyber attack on a hospital would breach other law of armed conflict rules, if the view is taken that such a cyber weapon is indeed a booby trap, this will have the effect that it would be unlawful for states party to Protocol II and/or Amended Protocol II to use such a cyber weapon if it is in any way associated with any of the items listed in article 6(2) of Protocol II or article 7(1) of Amended Protocol II.

[92]As an alternative example, consider a similar operation that takes control of a RPA, this time carrying an incendiary weapon within the Protocol III definition. If the State taking control is party to Protocol III, any use of the incendiary weapon against its enemy must comply with that treaty, particularly article 2.

in the definitions in a number of weapons law treaties. Imagine, for example, that it becomes possible to take cyber control of a laser system originally designed as a range finder and then to re-design by cyber means the way in which the laser operates so as to give it a combat function to cause permanent blindness to unenhanced vision. Such a re-designed weapon would be prohibited by article 1 of Protocol IV, if, that is, the remaining elements of that article are satisfied. Accordingly, a State undertaking such a cyber re-designing operation must have regard to the law by which it is bound in deciding whether the re-designed weapon, or the associated method of warfare, is the subject of a prohibition or restriction by which the relevant State is bound.

7 Conclusions

This paper has shown that notions of remote attack generate ethical concerns that have been discussed since the times of the ancient Greeks. Military commanders since time immemorial have sought to attack their enemies from a safe distance and thus to place the enemy at risk while preserving their own security. It remains to be seen whether emerging forms of remote attack are merely seen as the most recent manifestations of this trend, or are interpreted as requiring the development of new, ad hoc legal rules.

Any state acquiring or adopting a new weapon is under a legal obligation to conduct a weapon review. Existing rules of weapons law provide the benchmark against which the acceptability of the new weapon is to be judged. States party to API are additionally required to review new methods of warfare and to conduct these reviews when new weapons, means or methods are being studied or developed. The principles and rules of weapons law have been summarised in the present paper and we have considered three types of remote attack in turn, namely RPAs, autonomous highly automated weapon systems, and cyber weapons. We have applied the principles and rules of weapons law to each of these. We have also noted that in the case of weapon systems that take human beings out of the loop of attack decision-making, a weapon reviewer will have to assess whether the weapon system, in its intended manner of use, will enable the rules of targeting law to be complied with.

A general conclusion that emerges is that weapons law and the law of targeting seem to be robust enough to provide a sensible basis for differentiating between remote attack systems that are acceptable and those that are not. Autonomous offensive attack technology such as is currently known to exist cannot be used in compliance with targeting law, for example because it is incapable of undertaking the evaluative decision-making required by certain of the precautionary rules in article 57 of API. While such a weapon system based on currently known technology should therefore fail a weapon review, it would seem to be premature for States to ban autonomous attack technologies in general. Scientific and technical development of autonomous systems may, in the future, yield opportunities for improved compliance with distinction, discrimination, proportionality and precautions, and it would be most unfortunate if a blanket ban were to preclude the possibility of realising those opportunities.

References

Akerson D (2013) The illegality of offensive lethal autonomy. In: Saxon D (ed) International humanitarian law and the changing technology of war. Nijhoff, Leiden, pp 65–98

Anderson K, Waxman M (2013) Law and ethics for autonomous weapon systems: why a ban won't work and how the laws of war can. Hoover Institution, Stanford University. //papers.ssrn.com/sol3/papers.cfm?abstract_id=2250126. Accessed 13 Dec 2015

Backstrom A, Henderson I (2012) New capabilities in warfare: an overview of contemporary technological developments and the associated legal and engineering issues in Article 36 weapons reviews. IRRC 94:483–514

Boothby WH (2009) Weapons and the law of armed conflict. Oxford University Press, Oxford

Broad WJ, Markoff J, Sanger DE (15 Jan 2011) Israeli test on worm called crucial in Iran nuclear delay. In: The New York Times. http://www.cfr.org/iran/nyt-israeli-test-worm-called-crucial-iran-nuclear-delay/p23850. Accessed 13 Dec 2016

The Daily Telegraph (25 Feb 2013) Israel stages test flight of arrow 3 missile defence. http://www.telegraph.co.uk/news/worldnews/middleeast/israel/9893150/Israel-stages-test-flight-of-Arrow-3-missile-defence.html. Accessed 13 Dec 2016

Daoust I, Coupland R, Ishoey R (2002) New wars, new weapons? The obligation of states to assess the legality of means and methods of warfare. IRRC 846:345–363

Dinstein Y (2010) The conduct of hostilities under the law of international armed conflict, 2nd edn. Cambridge University Press, Cambridge

Droege C (2012) Get off my cloud: cyber warfare, international humanitarian law, and the protection of civilians. IRRC 94:533–578

Fenrick WJ (1990) The conventional weapons convention: a modest but useful treaty. IRRC 279:498–509

Fidler DP (2005) The meaning of Moscow: 'Non-Lethal' weapons and international law in the 21st century. IRRC 87:525–552

Fildes J (23 Sept 2010) Stuxnet worm attacked high value Iranian assets. In: BBC News. http://www.bbc.co.uk/news/technology-11388018. Accessed 13 Dec 2015

Haines S (2012) The nature of war and the character of contemporary armed conflict. In: Wilmshurst E (ed) International law and the classification of conflicts. Oxford University Press, Oxford, pp 9, 11

Hays Parks W (2005) Conventional weapons and weapon reviews. YIHL 8:55–142. https://doi.org/10.1017/S1389135905000553

Hays Parks W (2006) Means and methods of warfare, symposium in honour of Edward R Cummings. GWIR 38:511–542

Henckaerts J-M, Doswald-Beck L (eds) (2004) Customary international humanitarian law, vol. I: rules. Cambridge University Press, Cambridge

Homer, Illiad, 13.262-3

Lewis M et al (2009) Scaling up wide-area-search munition teams. IEEE Intell Syst 24(3):10–13. https://doi.org/10.1109/MIS.2009.63

McClelland J (2003) The review of weapons in accordance with Article 36 of additional protocol 1. IRRC 85(850):397–420

O'Connell RL (1989) Of arms and men: a history of war, weapons and aggression. Oxford Paperbacks, Oxford

Oeter S (2013) Methods and means of combat. In: Fleck D (ed) The handbook of international humanitarian law, 3rd edn. Oxford University Press, Oxford, pp 115–230
Program on Humanitarian Policy and Conflict Research (HPCR) (2010) The HPCR manual on international law applicable to air and missile warfare (AMW Manual). Harvard University, Cambridge University Press, Cambridge
Rapporteur's Report CDDH/215/Rev.1 paragraph 27 reported in ICRC Commentary, paragraph 1454
Richmond J (2012) Evolving battlefields: does STUXNET demonstrate a need for modifications in the law of armed conflict? Fordham Int Law J 35:842–894
Rowe PJ (1987) Defence. The legal implications. Brassey's Defence Publ., London
Schachtman N (28 Sept 2011) Army tracking plan: drones that never forget a face, available at http://www.wired.com/2011/09/drones-never-forget-a-face/. Accessed 13 Dec 2016
Schmitt MN (2011) Cyber operations and the jus in bello: key issues. Int Law Stud 87:89–110
Schmitt MN, Thurnher J (2013) 'Out of the loop': autonomous weapon systems and the law of armed conflict. Harv Natl Secur J 4:231–281
Schmitt MN (ed) (2013) Tallinn manual on the international law applicable to cyberwarfare. Cambridge University Press, Cambridge
Scobbie I (2012) Gaza. In: Wilmshurst E (ed) International law and the classification of conflicts. Oxford University Press, Oxford, pp 280–316
Singer PW (2011) Robots at war: the new battlefield. In: Strachan H, Scheipers S (eds) The changing character of war. Oxford University Press, Oxford, pp 333–356
Solis GD (2011) The law of armed conflict. Cambridge University Press, Cambridge
Sparrow R (2007) Killer robots. J Appl Philos 24(1):62–77. https://doi.org/10.1111/j.1468-5930.2007.00346.x
UK Ministry of Defence (13 Nov 2007) DCDC, Future maritime operational concept 2007
UK Ministry of Defence (2010) Legal support to joint operations. Ministry of Defence Development, Concepts and Doctrine Centre, JDP 3-46 dated August 2010
UK Ministry of Defence (2004) Manual on the law of armed conflict (UK Manual)
University of Birmingham Policy Commission Report (Oct 2014) The security impact of drones: challenges and opportunities for the UK. http://www.birmingham.ac.uk/Documents/research/policycommission/remote-warfare/final-report-october-2014.pdf. Accessed 13 Dec 2016
Wagner M (2011) Taking humans out of the loop: implications for international humanitarian law. J Law Inform Sci 11:1–11

Part II
Ethical Challenges of Dehumanization

Dehumanization: The Ethical Perspective

Jan C. Joerden

Abstract To structure the debate on the ethical admissibility of "dehumanization" of warfare, e.g. by using drones, the distinction between evaluation rules and imputation rules seems to be helpful. Evaluation rules give information as to whether a certain behavior is obligatory, forbidden or permitted. From these rules, the rules that give information as to whether or not a person can actually be held responsible for a certain behavior must be distinguished; for this requires imputation. And without imputation of a behavior as an act (or omission) of a certain person, it is not even reasonable to apply the evaluation rules to this behavior. For it is by applying the imputation rules that the *subject* is determined to which the evaluation rules can then refer to. This also means that there are two categorically different avenues to shed light onto "dehumanized" behavior in war: It is possible that there are evaluation rules that forbid certain "dehumanized" behavior as a matter of principle; and it is possible that there are imputation rules, the violation of which could lead to persons successfully but unjustifiedly evading their own responsibility. The first avenue will be—after some introductory remarks in sections "Two Meanings of 'Dehumanization'" to "The Distinction Between Evaluation Rules and Imputation Rules"—examined below in sections "Ban on Usage of Certain Weapons and Weapon-Systems" to "Drones and 'Stealth'", while the second is then investigated in sections "The Distinction Between Combatants and Civilians" and "Problems of Imputation".

Jan C. Joerden holds the Chair of Criminal Law, especially International Criminal Law and Comparative Criminal Law, Legal Philosophy, European University Viadrina Frankfurt (Oder), Germany. For the translation of this paper I have to thank my assistant Dr. Johannes Bochmann, Frankfurt (Oder)/Berlin.

J.C. Joerden (✉)
Lehrstuhl für Strafrecht und Rechtsphilosophie, Europa-Universität Viadrina,
Große Scharrnstraße 59, 15230 Frankfurt (Oder), Germany
e-mail: joerden@europa-uni.de

W. Heintschel von Heinegg et al. (eds.), *Dehumanization of Warfare*,
DOI 10.1007/978-3-319-67266-3_4

1 Two Meanings of "Dehumanization"

The term "dehumanization" has at least the following two meanings. Firstly, it can characterize a tendency referring to war victims' sufferings. If, for example, particularly cruel or perfidious weapons are used, this can be referred to as "dehumanization" of war. International Humanitarian Law aims at countering this kind of "dehumanization".

On the other hand, there is a meaning of the term "dehumanization" that must be set apart, namely, that mankind withdraws, so-to-speak, from the immediate war-related events and thus, "dehumanization" of war occurs. Ideally, one can imagine that only pieces of technical equipment fight against one another, for example in form of a purely virtual military conflict on the Internet or as a sort of "Star Wars" in space without direct contact with the enemy. The war would then be decided once the enemy's military potential has been destroyed; the population would be spared.

That an abolition of man-killing wars through such a "dehumanized" scenario is rather illusive, is perfectly clear, because man can hardly imagine these kind of "proxy wars", in which no blood flows, as sufficient to fight out conflicts of interest. As little as the few heroic examples in which the two leaders of hostile peoples only fought each other in order to spare their peoples have gained acceptance in the end. The peoples will not agree, not least because they will only be prepared to acknowledge that they have lost when they themselves have lost, and not already when only their leader has lost.

2 Unfairness of Asymmetric Warfare?

It is conceivable, however, and also put into practice already by using unmanned aerial vehicles such as drones,[1] but also rockets etc., that man withdraws from the immediate front-line, by using weapon-systems that can be controlled from a distance or that can even control themselves (the so-called "autonomous systems"). Yet this approach is only truly attractive tactically and strategically if the enemy does not have comparable weapon-systems or sufficient means to counter such attacks; rather, a considerable technical advantage is necessary, because otherwise a stand-off situation occurs that is known from the atomic "overkill" logic during the Cold War: neither of the two superpowers could afford direct confrontation with the other superpower, because, while killing the defending power, this would at the same time have amounted to suicide of the attacking power.

The usage of remote-controlled or self-guided and self-controlling weapon-systems[2] is therefore only strategically and tactically reasonable in cases of

[1]For the legal and ethical problems with using drones cf. especially Frau (2014a).

[2]For details of the technical realization cf. Mahn-Gauseweg (2014).

technological superiority or some other superiority, for example air supremacy over a certain area etc. Put otherwise, the usage of said weapon-systems is always a manifestation of a kind of asymmetric warfare: the enemy is attacked with means or methods that he does not have available himself. Not by chance are such weapon-systems also especially used as an answer to an enemy's warfare that is asymmetric in other respects, for example in the so-called War on Terrorism. Here, too, the enemy uses means and methods that the attacked party does not have available, namely selective terror that is brought to the civilian population, and therefore a war without clear front-lines and without the self-restrictions by the rules of International Humanitarian Law.

Despite this quasi symmetric asymmetry of warfare, for example of the USA against the Taliban, the question must be asked, whether waging such an asymmetric war is ethically acceptable at all. And this question must even be asked if it were clear what party is justified to wage war and what party is not. Even if a justified (defensive) war is fought, from an ethical point of view, this does not justify the usage of *all* methods of warfare. This is displayed not least by International Humanitarian Law, that is undoubtedly based on ethical considerations at its core. It could therefore well be the case, that asymmetric warfare is ethically unacceptable even if the purpose of warfare may be justified, simply because the (good) end does not justify the (bad) means. Insofar, fighting asymmetric wars could always be unfair. Fairness, meanwhile, is admittedly an ethical category, though not one that acts of war can reasonably measured against. The possibility of waging a fair war that might have been called "knightly" in earlier days,[3] although it will hardly ever have deserved this positively connoted attribute, is illusive long since. A war is not a duel in which chances may in general be relatively fairly distributed, and—at least in theory—it only depends on the duelists' personal skills. To continue with this metaphor, the ethics of warfare does not prohibit the usage of pistols in a duel against an enemy even if he is equipped only with a knife. To fight asymmetric (defensive) wars is, therefore, not unethical *per se*.

However, "dehumanization" of war may be unethical for another reason, namely because whoever uses remote-controlled weapon-systems is located outside the "danger zone" of war and, to some extent like in a computer game, merely operates the systems without even coming close to danger himself.[4] As far as this, once again, resonates the allegation of lack of "knightliness", this allegation is arguably as unsustainable as the unfair warfare qua asymmetry just mentioned. For who would seriously blame a war-making party for protecting their own soldiers as ideally as possible from endangerment by the enemy. This is the notion that is not unethical *per se* that is also the basis for the usage of tanks or jet fighters flying above the enemy radar or other self-protective measures.

---

[3]Cf. e. g. Leveringhaus and Giacca (2014), 22 et seq.; Borrmann (2014), 172 et seq.

[4]Cf. Banaszewska (2014).

3 The Distinction Between Evaluation Rules and Imputation Rules

In order to bring some structure to the debate on ethical admissibility of "dehumanization" of warfare, e. g. by using drones, I would like to point to the following differentiation that is essential for judging a person's behavior (part of which can be behavior in a "dehumanized" war): The distinction between evaluation rules (*Bewertungsregeln*) and imputation rules (*Zurechnungsregeln*).[5] Evaluation rules give information as to whether a certain behavior is obligatory, forbidden or permitted. For example, the killing of another person (without grounds of justification) is forbidden. Even if killing other persons (enemies) is permitted, or even obligatory, the same applies structurally, because according to the prevailing view the general prohibition to kill is invalidated by a ground of justification, namely waging (justified) war. In the same way as prohibition rules, the obligation rules and permit rules belong to this group of evaluation rules.

From these evaluation rules, the rules that give information as to whether or not a person can actually be held responsible for a certain behavior must be distinguished; for this requires imputation. And without imputation of a behavior as an act (or omission) of a certain person, it is not even reasonable to apply the evaluation rules to this behavior. For it is by applying the imputation rules that the *subject* is determined to which the evaluation rules can then refer to. Put otherwise: Before an act of killing can be criticized as illegal or immoral, it must be clear that the events taken into consideration and possibly criticized are, in fact, an act of killing.

This, however, also means, that there are two categorically different avenues to shed light onto "dehumanized" behavior in war: It is possible that there are evaluation rules that forbid certain "dehumanized" behavior as a matter of principle; and it is possible that there are imputation rules, the violation of which could lead to persons successfully but unjustifiably evading their own responsibility.

The first avenue will be examined below in chapters "Ban on Usage of Certain Weapons and Weapon-Systems" to "Drones and 'Stealth'",[6] while the second is then investigated in chapters "The Distinction Between Combatants and Civilians" and "Problems of Imputation".

---

[5]Cf. Hruschka (1987, 1991); Joerden (2015).

[6]These chapters contain some revised parts of Joerden (2014).

4 Ban on Usage of Certain Weapons and Weapon-Systems

International Law recognizes the fact that not every means that are suitable for waging war are also permitted in war.[7] This can be seen, among others, by the ban on certain weapons, for example cluster bombs, and particularly chemical and biological warfare agents.[8] As far as there are binding rules of International Law, at least the legal situation is clear, of course at the same time without it being ascertained that every State in fact abides by these rules; however, in this regard the situation in Public International Law is not different from the one in other areas of Law, in particular Criminal Law, that is also not able to guarantee that no one acts against Criminal Law.

Increasingly, new weapon-systems are now being developed; for example the above-mentioned unmanned drones, with which other countries can be overflown without putting one's own soldiers at risk, and that are suitable, if constructed accordingly, not only for mere surveillance but also to fire guided weapons that are then directed to an enemy target and, among others, kill people there (the additional possibility of causing damage to property shall be left out of consideration in the following). Some of these drones can even make the "decision" to fire guided weapons "autonomously" on the basis of an electronic processing of data from the environment and then set in motion the weapons carried along. Other drones are operated from distance by accordingly trained soldiers, who then must also approve of the activation and firing of weapons.[9]

For these new weapon-systems, the question arises—even before the possibility of a corresponding ban by International Law—whether they should be labeled as damnable, similar to cluster bombs or chemical or biological weapons, which could lead to their ban by International Law after respective negotiations between the States. For such a label as damnable, however, an appropriate criterion is needed. For the fact alone, that people can be killed with the aid of them is (at least for the time being) not sufficient for a ban on weapons, or their usage, under International Law.[10]

[7]This refers to the so-called ius in bello, cf. The Hague Convention respecting the Laws and Customs of War on Land of 1907, the Convention on the Prevention and Punishment of the Crime of Genocide of 1948, the Geneva Conventions of 1949, and the Rome Statute of the International Criminal Court of 1998, to name just the most important attempts to limit waging war. These rules are supplemented by further Conventions to prepare or update these rules, as well as by Customary International Law, cf. in greater detail, Werle and Jeßberger (2014) marginal nos. 151 et seq., 1030 et seq.

[8]In contrast, a spanning ban on nuclear weapons cannot be found either in International Treaty Law or in Customary International Law; for more details cf. Werle and Jeßberger (2014), marginal no. 1400.

[9]More information on this e. g. in Leveringhaus and Giacca (2014).

[10]It will most likely always remain a pacifistic ideal to ban all weapons suitable for warfare and lethal to humans in order to thus make wars impossible. It is another matter, however, whether this goal should be maintained at least as a moral imperative, with the consequence, inter alia, that

5 Kant's Suggestion

It is well-known that there has been a rather prominent attempt in the history of philosophy to set up a general criterion for the question what means of warfare may not be used under any circumstances. In his work *Towards Perpetual Peace* from 1795, Immanuel Kant phrased to following demand[11]: "No state during a war is to permit acts of hostility that would make mutual confidence impossible after the war is over – e.g. the use of assassins (*percussores*) and poisoners (*venefici*), breach of capitulation, incitement to treason (*perduellio*) in the opposing state". Kant gives a reason for this demand by saying that these are "dishonorable strategems."[12] "Even in war there must be *some* confidence in the enemy's character: otherwise no peace could be concluded, and the hostilities would degenerate into a war of extermination (*bellum internecinum*)".[13] Kant points out that war is merely a "sad recourse by which each state uses violence to assert its right". Because there is no "higher court to settle disputes through Law" in the "state of nature" between two States, none of the war-making parties "can be condemned as wrong, because that would presuppose a judicial decision."[14]

Thus, in the end the outcome of a war determines the answer to the question which side is right, "as though this were a judgment of God"[15]. And Kant adds that a war to punish (*bellum punitivum*) between States is unthought of, because "no State has authority over any other State" (such authority, however, is a precondition for punishment). Kant draws the conclusion that a war of extermination which can wipe out both parties and all justice, can lead to 'perpetual peace' only in the vast burial ground of the human race. Such a war, therefore, "must be absolutely forbidden."[16]

That the mentioned means of warfare will "inevitably" lead to such a war of extermination, is a conclusion Kant draws from the idea that "once these vile practices [literally: hellish arts] are employed, they'll soon spread beyond the confines of war, ... and will carry this over into the state of peace and thereby cancel the spirit of peace" completely.[17] As an additional example for these vile practices of warfare, Kant adds the "use of spies" (*uti exploratoribus*), which "only makes use of the infamy of others (something that can't be entirely exterminated)".

In summary Kant's position can be characterized by saying that he rejects all means of warfare that operate against the enemy using particular perfidiousness[18]

---

waging war, provided that innocent civilians are affected—which is generally unavoidable, can be possibly excused but by no means be justified morally.

[11]6th preliminary article towards perpetual peace, Kant (1795).

[12]Ibid.

[13]Ibid.

[14]Ibid.

[15]Ibid.

[16]Ibid.

[17]Ibid.

[18]The Latin term "perfidus" can mean, among others, "faithless", "dishonest", or "treacherous".

that goes beyond the mere killing, the latter being unavoidable in war. These are means—in general connected with deceit—that destroy the possibility of rebuilding confidence or trust towards a person who uses them, because such a person has, so to speak, shown that he does not abide by any ethical boundaries, and therefore is not a potential partner for concluding perpetual peace—even after a war ended. Whosoever uses assassins[19] or poisoners,[20] does not comply with the capitulation terms,[21] or incites to treason, displays himself to be dishonorable and figuratively extinguishes even the slightest light at the end of a tunnel of war, that at least hopes for peace.

It becomes clear that, regarding his prohibition rules, Kant places special emphasis on those means of warfare that are connected to—in a broad sense—fraudulent intrigues. For a start, this may have to do with Kant's disgust for any form of lies, that he expressed in particular in his essay "On the Supposed Right to Lie From Benevolent Motives",[22] which he wrote in 1797, i.e. two years after *Perpetual Peace*. In this essay, Kant denies the possibility of a Right to Lie, even if this lie could serve to rescue a person from his persecutors. In the context of *Perpetual Peace*, the examples Kant chooses, however, rather seem to be assigned the "principle of publicity"[23] (*Princip der Publicität*) which Kant introduces in precisely this writing. According to this principle, the "transcendental formula of public law" is: "All actions relating to the right of other men are unjust if their maxim is not consistent with publicity."[24] The details of this formula and its application, for which Kant does give some examples, cannot be illuminated further in this context.[25] Meanwhile, it should be quite clear that the "vile practices" ["hellish arts"] of warfare, as Kant calls them, do not harmonize with the "principle of publicity". For would you make known the intended objectives pursued by those means beforehand, they would be foredoomed to failure. The assassin who announces his intentions, the poisoner, the capitulation breaker, the agitator to treason, and also the spy must fear the light of publicity if they want to succeed. Their actions are incompatible with the "principle of publicity".

It is striking, that among the means of warfare that ought to be banned, Kant does not mention any weapons or their use that are problematic due to their specific principle of operation. This may especially have to do with the fact that Kant could not know anything about mines, cluster bombs, or fragmentary bombs, much less

[19] Cf. sec. 11 para. 1 no. 7 German Code on International Criminal Law (Völkerstrafgesetzbuch - dVStGB).

[20] Cf. sec. 12 para. 1 no. 1 dVStGB.

[21] "Breach of capitulation" arguably includes both cases in which someone takes action against a person who has already capitulated (sec. 8 para. 1 no. 1 in conjunction with para. 6 no. 3 dVStGB), as well as cases in which someone only feigns his capitulation in order to be able to attack the other party more effectively.

[22] Kant (1797).

[23] Kant (1795), Appendix II.

[24] Ibid.

[25] Cf. in greater detail Joerden (2011) with further references.

about biological or chemical agents of war. But undoubtedly even in his time there were already weapons that, by their method of killing, were able to cause more suffering and torments than necessary for the act of killing as such. Likewise, the use of torture to extract military relevant information, a means of warfare in a broader sense which is undoubtedly outlawed around the world[26] nowadays, was known. That Kant did not want to morally and legally thrust aside these and other means of warfare (acting quasi physically) in the same way, but rather restricted himself to cases of application of the "principle of publicity", may also be connected to the fact that he believed it was highly unrealistic to prohibit such common means of warfare with a general ban (by International Law). Today, under the impression of all kinds of weapons of war that work rather brutally, the deception-based means of warfare Kant wanted to ban have begun to disappear from view.[27] It remains, that Kant was one of the first who attempted to formulate ethical arguments for marking (international) legal limits of warfare and thus outlawing certain means of warfare.

6 The Difference Between Murder and Intentional Homicide in German Criminal Law as Distinctive Criterion

If in the following "simple" intentional killings shall be distinguished from "aggravated" (i.e. particularly damnable) intentional killings in war times, I would suggest that some use can be made of the demarcation between intentional homicide (sec. 212 German Criminal Code [*Strafgesetzbuch* – dStGB]) on the one hand and murder (sec. 211 dStGB) on the other hand.[28] Even though this concept is, so to speak, designed only for peace times, it nonetheless reflects the attempt to

[26]Cf. the UN Convention against Torture and Other Cruel, Inhuman or Degrading Treatment or Punishment, cf. also sec. 7 para. 1 no. 5 dVStGB. – To what extent it could be permissive, exceptionally, to use torture in peace times to save human lives is highly controversial, at least in Germany; cf. the references on this debate, e.g. by Lackner and Kühl (2014), § 32 marginal no. 17a, and in more detail with further references Joerden (2005).

[27]Nonetheless, some prohibitions directed at these means can be found in the German Code of International Criminal Law (dVStGB); cf., e.g. sec. 11 para. 1 no. 7 dVStGB on the "assassin", sec. 12 para 1 nos. 1 and 2 on the "poisoner", sec. 8 para 1 no. 1 in conjunction with para. 6 no. 3 dVStGB on "breach of capitulation", sec. 8 para. 3 no. 4 dVStGB on "incitement to treason". This, however, shall not imply that there is complete congruence between the respective "vile practices" ["hellish arts"] Kant names on the one hand and the actions covered by the quoted provisions of the dVStGB on the other hand.

[28]One might ask whether the regulations of the Rome Statute or the mostly corresponding German Code of International Criminal Law (dVStGB) are better suited to guide such a distinction. However, these rules approach the problems rather casuistically and thus hardly contain general criteria for the distinction at hand here. Rather, particularly the rules in the dVStGB for their part, are based on the distinction between intentional homicide (sec. 212 dStGB) and murder (sec. 211 dStGB), including the rule on genocide [literally "peoples' murder" in German] in sec. 220a dStGB that was abolished at the same time the dVStGB was implemented.

qualify particularly damnable killings as murder and thus distinguish them from other killings that are not as damnable and are referred to as intentional homicide. This is the same goal that is pursued when considering a ban of certain especially damnable means of warfare. Intentional homicide in the German Criminal Code is characterized by the fact that the offender "kills a person without being a murderer under sec. 211" (sec. 212 dStGB). On the other hand, murder requires that the offender "kills a person for pleasure, for sexual gratification, out of greed, or otherwise base motives, by stealth, cruelly, by means that pose a danger to the public, or in order to facilitate or to cover up another offence" (sec. 211 para. 2 dStGB).[29]

Regarding the methods of warfare that ought to be banned in wars, there is apparently also a differentiation between "simple" killings that can generally be justified by International Law as "normal" appearances of warfare, at least as long as it actually is a "just(ified)" war altogether and not, e.g., an unjust(ified) war of aggression.[30] At this point, however, I must work on basis of the—admittedly controvertible[31]—thesis that people may be killed in a war, because the state of war as such, at least if it can be deemed to be justified on part of the acting party, functions as ground of justification regarding the killing of an enemy[32] (as far as the lethal force is legally authorized by a person on the relevant level of the command structure, is not disproportionate, etc.). For the sake of argument, I must presume that there actually are justified wars at all. For all these international legal preliminary questions must already be answered positively, before it comes to differentiating between acceptable and unacceptable use of weapons. Therefore, if all this is assumed, the difference in German Criminal Law between intentional

[29]It is perfectly clear that the wordings of secs. 211, 212 dStGB refer to peace times and rather have the single perpetrator in mind. Anyway, this article is not so much concerned with criminal proceedings against an offender, but rather with the underlying ethical differentiation between "simple killings" and "particularly damnable killings", i.e. murder. – One also might ask whether the problematic wording of sec. 211 dStGB, of all things, can be helpful for the differentiation discussed here. For firstly, it is common knowledge that the legislative history of sec. 211 dStGB is very problematic because this regulation was implemented in 1935 (cf., e.g. Wolf [1996]) and signalized a departure from Criminal Law that focused on the offense ("Tatstrafrecht") and an orientation towards the idea of Criminal Law being focused on the offender ("Täterstrafrecht"). Additionally, the borderline between murder and intentional homicide by law is indeed debated; cf. in recent times on respective reform efforts Kubik and Zimmermann (2013) with further references. Even given this critique of secs. 211 and 212 dStGB and their systematic relation to each other, it still seems that the current wording of sec. 211 dStGB includes many (whether all, I will leave unanswered at this point) forms of action that would be characterized as particularly damnable in the context of war, too; cf. hereinafter.

[30]As far as there is no ius ad bellum, at least according to more recent perceptions on non-justifiable war of aggression, waging war that is not defensive is a violation of (Public International) Law, in fact even if this war complies with the rules of ius in bello.

[31]A pacifist, on the other hand, would be able to hold the view that the killing of an enemy can never be justified, but rather excused, at most.

[32]Whereat the killing of civilians is excluded here; and even combatants may only be killed if they have not surrendered etc.

homicide and murder can at least become an orientation guide when it comes to distinguishing between acts of killing in war that are yet acceptable and those that are not acceptable anymore.[33]

For even in times of peace it appears legally problematic to justify murder at all.[34] To justify the active killing of another person, actually only a situation of self-defense comes into consideration, i.e. a situation, in which an immanently unlawful attack against a legally protected interest is averted by use of the necessary and demanded [*geboten*][35] defensive measures (cf. sec. 32 dStGB). However, the problems arising here is consistently covered by the fact that, when it comes to the possible justification of murder—according to the prevailing view in opinion in German legal scholarship that presumes the relationship between intentional homicide and murder to be one of a basic offense [*Grundtatbestand*] (sec. 212 dStGB) and a qualified offense [*Qualifikation*] (sec. 211 dStGB)—first asks whether or not the intentional homicide is justified and, if this is the case, does not consider murder and its justification at all anymore, because there can be no qualified offense if the basic offense is not committed unlawfully: If the basic offense is not committed unlawfully, the substantive elements of the qualified offense cannot be fulfilled.

Those who follow the opinion voiced in constant jurisdiction, namely, that the crime of murder according to sec. 211 dStGB is a "*delictum sui generis*", must raise the relevant questions of justification by self-defense when considering the crime of murder and, if sec. 32 dStGB does apply, deny that a chargeable act of killing exists; for intentional homicide according to sec. 212 dStGB must not be examined anymore because the relevant category of offense—if one or more murder criteria are fulfilled—is only murder according to sec. 211 dStGB from the start.[36] In contradiction to the prevailing scholarly opinion, the question of justifying the killing of a person while fulfilling one or more murder criteria comes into view of a possible justification immediately. In a way, the whole offense of murder must be "refuted" by justification.

However the exact relationship between intentional homicide and murder must be described accurately, even the prevailing scholarly view that distinguishes insofar

---

[33]This is not meant to imply a thesis that there are no other acts besides those (indirectly) included in sec. 211 dStGB which should be banned or are already prohibited by the Rome Statute, other rules of International Law or the dVStGB. It rather seems to be that at least those kinds of killing covered by sec. 211 dStGB are particularly damnable and thus cannot be justified in war, too.

[34]Cf. Löffelmann (2013), p. 500; Eser and Sternberg-Lieben (2014) § 211 marginal no. 41.

[35]The term "demanded" ("geboten") derives from the wording of sec. 32 dStGB and makes clear, according to the prevailing opinion, that so-called social-ethical limitations of the right to self-defense must be considered, e.g. in cases of provocation of self-defense, flagrant disproportionality between the attacked interest and the attackers interest that is affected by the defense, and in cases of self-defense against persons belonging to certain groups (in particular not criminally responsible persons).

[36]The question whether the BGH (Bundesgerichtshof – Federal Court of Justice) would actually draw this consequence in every relevant case shall remain unanswered at this point.

between a basic offense and a qualified offense, in the end does not get around including the circumstances of fulfilling murder criteria into the evaluation of a potentially justifying situation when examining the possible justification by self-defense according to sec. 32 dStGB. For the fulfillment of murder criteria could also raise the questions of justification of the underlying intentional homicide by self-defense. However, a differentiated view on the murder criteria of sec. 211 para. 2 dStGB is necessary, namely depending on whether they are "offender-related" criteria, i.e. criteria dealing with motives or specific intents,[37] or "offense-related" criteria.[38] Fulfilling offense-related murder criteria could at least effect the attribute of necessity of self-defense, possibly also (only) the attribute of "demandedness". On the other hand, fulfilling offender-related intent or motive criteria could question the so-called "will to defend" that is necessary for justification as an act of self-defense. At this place, this cannot be exemplified for each and every murder criterion.[39] But at least for the offense-related murder criterion of "stealth" [*Heimtücke*], I will here attempt to make the thesis just mentioned plausible.

Whoever kills "by stealth" takes advantage (with hostile will[40]) of his victim's artlessness and defenselessness; this is at least the overwhelmingly prevailing interpretation of "by stealth" [*Heimtücke*] in Germany.[41] Now, how can a case in which a murderer by stealth can be justified by self-defense even be thought of? In order for this to become possible, the artless and defenseless person would have to be in the process of performing an imminent, unlawful attack, which does appear contradictory. Because whoever is imminently unlawfully attacking another person will hardly be artless and defenseless regarding the attacked person's[42] counter reaction.[43] So, therefore, if the substantial elements of murder by stealth are indeed fulfilled, this murder cannot be justified as an act of self-defense. Even in the highly debated cases in which a suppressed and smitten family member kills a sleeping

---

[37] 1st and 3rd group of murder criteria in sec. 211 para. 2 dStGB: pleasure, sexual gratification, greed, otherwise base motives, intent to facilitate or cover up another offense.

[38] 2nd group of murder criteria: stealth, cruelty, committal by means that pose a danger to the public.

[39] Cf. in greater detail in Joerden (2014).

[40] This is a condition that judicature establishes; the probably prevailing opinion in legal scholarship differs and requires a particularly damnable breach of trust by the offender in addition.

[41] Cf. Lackner and Kühl (2014), § 211 marginal no. 6.

[42] The attacker's artlessness and defenselessness will also no longer exist regarding (possible) reactions of a third party acting to defend the attacked person [Notwehrhelfer; note that self-defense in German Criminal Law includes the defense of others], so that the third party, too, does not commit a murder by stealth which is justified as an act of self-defense but merely a justified intentional homicide. For whosoever attacks another person is no longer artless not only regarding this person's possible defense but also regarding possible defensive acts by any third party, exactly because he knows that he is acting illegally and thus must and will always take into account possible counter-reactions by persons that want to defend the attacked person or the legal order.

[43] Even in cases that appear somewhat constructed, in which B, who only pretends to be sleeping, seems to be asleep and is attacked by A, one can hardly say that A is artless and defenseless, if the seemingly sleeping B "awakens" and stabs the attacker to death.

household tyrant,[44] (arguably) by stealth,[45] as far as I can see, no-one assumes such a killing to be justified as an act of self-defense, at least because the defense is not necessary, to be more precise: because the attack is not imminent: Whoever is sleeping, is not attacking.[46]

7 Consequences for Evaluating Certain Acts of Killing in War

Now one might argue that acts of killing can generally be justified in war times—at least as long as the above-mentioned preconditions are met. But that at the same time—just as in peace times—the possibility of justification of acts of killing is limited legally and ethically, at least in cases in which murder criteria are fulfilled at the same time.[47] Looking at the murder criteria listed in sec. 211 dStGB from this point of view, it attracts attention that the use of specific weapons that are banned by the international community might well be subsumed under the term "cruel" [*grausam*] commission of the crime. For these weapons, i.e. specific ammunition, chemical and biological agents of war, etc.[48] stand out negatively because they inflict far more pain and torment on their victims than would be necessary for the act of killing itself.[49] This should be avoided in any event and therefore these weapons and weapon-systems are banned by International Law.

This parallel seems to be less obvious when dealing with murder using "means that cause a danger to the public". For most weapons used in war pose a danger to the public; think only of weapons used by the artillery or the air force, the use of which is not only directed against the attacking enemy but regularly also

[44]According to the probably prevailing scholarly opinion the house tyrant took his "artlessness" into his sleep; cf. on the killing of sleeping persons as murder by stealth, e.g. Lackner and Kühl (2014), § 211 marginal no. 7 with further references, including judicature.

[45]Of course, there is legitimate subject for debate as to whether this is in fact a case of a murder by stealth. If, in accordance with the probably prevailing scholarly opinion, a "particularly damnable breach of trust" is seen as an additional requirement of "stealth", this may well not be the case.

[46]A justification under the (general) rules on a defensive state of emergency (analogous application of sec. 228 dBGB) will also mostly fail due to the criterion of "necessity" to ward off the danger, because the rule "Justice does not have to yield to injustice" [Recht braucht dem Unrecht nicht zu weichen; cf. the castle doctrine in US law, according to which there is no duty to retreat] does not apply in the context of a defensive state of emergency. – Apart from that, only a ground of excuse could come into question, cf. the references on the respective debate, given, e.g. by Lackner and Kühl (2014), § 34 marginal no. 9.

[47]Indirectly, this finds expression when soldiers who are blamed that all soldiers are murderers defend themselves against this allegation, particularly because they do not intend to fulfill murder criteria and would also deny that every killing in war also constitutes murder.

[48]Cf. sec. 12 dVStGB.

[49]Apart from this, the aspect of endangering third parties, e.g. by splinter bombs etc. is regularly added.

against other persons that are not directly involved with the attack, including, in particular, the enemy State's civilian population. At this point, however, it becomes apparent that the problems are indeed comparable, as actions directed against the civilian population are in principle prohibited by (International) Law. Accordingly, for example sec. 7 para 1 no. 1[50] and sec. 11 para 1 no. 1 of the German Code on International Criminal Law (Völkerstrafgesetzbuch – dVStGB) penalize the intentional killing of members of the civilian population. Whoever uses weapons that also threaten members of the civilian population with death, at least comes under pressure to justify his behavior. To-date it has not been settled, whether or not the killing (and endangering) of civilians during war, often euphemistically referred to as "collateral damage", can ever be justified at all.[51] There is a strong indication that such killings may only be excused, at most.[52] This question cannot be investigated further at this point;[53] however, the fact still stands, that once again the murder criterion use of "means that pose a danger to the public" [*gemeingefährliche Mittel*] points towards a corresponding problem of justification when using such "means that pose a danger to the public" in war.

The same is true for the so-called motive or intention criteria of murder. Whoever performs an act of killing "for pleasure, for sexual gratification, out of greed, or otherwise base motives" will hardly be able to argue plausibly that he is carrying out a justified act of killing. Corresponding evaluations can also be found, inter alia, in the dVStGB that penalizes acts committed with intent to destroy a national, ethnical, racial, or religious group as a major crime (cf. sec. 7 para. 1 nos. 2 and 10 dVStGB). These crimes that are committed out of so-called racial hatred would have to be classed as murder out of "base motives"[54] according to the prevailing opinion. Even though there may not be a separate, distinct culpability for other acts of murder committed out of other motives found in sec. 211 dStGB, it does seem evident that the offender's motives are so condemnable that he cannot convincingly plead justification. Regardless of the question, whether such an offense committed in war will factually be prosecuted as a crime, there will hardly be any expert in International Law that will advocate the idea that killings committed out of such motives can be an acceptable part of a justified war. The same goes for acts of

[50]As a side-note it may be of interest that secs. 6 (genocide) and 7 (crimes against humanity) dVStGB can also apply outside of acts of war.

[51]Cf. in greater detail Merkel (2012).

[52]Dissenting, Merkel (2012), who considers an "impure" justification in the "normatively impure sphere of war" (p. 1137) referring to the concept of a "nonideal theory" debated by John Rawls.

[53]At least, the solution via the so-called duplex-effectus-doctrine, according to which collateral damages are seen as a quasi non-intended byproduct of the killing of combattants and are thus acceptable, that is often suggested is not convincing. In greater detail on possible conditions of applying the duplex-effectus-doctrine, particularly in Criminal Law, cf. Joerden (2007), p. 242 et seq.

[54]Cf. the references to Case Law, e.g., those given in Lackner and Kühl (2014), § 211 marginal no. 5a.

killing that are committed during a war in order to facilitate or to cover up another offense (cf. sec. 211 para. 2 group 3 dStGB).

Finally, the term "stealth" [*Heimtücke*], that also marks an act of killing as particularly damnable and thus as act of murder, will—as indicated above already—have to be subsumed under those means of war that inter alia Kant mentioned in his writing Towards Perpetual Peace. For the assassin, the poisoner, the breaker of capitulation, the agitator to treason, and the spy all act taking advantage of their victim's "artlessness and defenselessness". At least Kant deemed these means of war so damnable that he demanded a ban on them if "perpetual peace" should ever become possible.[55] Concluding, I would like to suggest a position on the evaluation of the use of drones against this background.

8 Drones and "Stealth"

The use of drones to collect information, i.e. for military reconnaissance, appears largely unproblematic in the present context,[56] simply because these actions are not directly connected to the killing of other persons and thus there is no parallel to a murder by stealth. Admittedly, such information can gain effectiveness regarding the killing of persons indirectly, however, this additionally requires an evaluation of the information gained as well as a human decision on the use of lethal means each time. This decision itself may be damnable, but it is not damnable solely because it is based on information collected using drones.

According to what was stated above, however, the use of drones to kill persons must generally risk being accused of amounting to a killing "by stealth". For both in cases in which a human decides and in which a drone "decides" itself, in order to kill, this weapon technology relies upon taking advantage of the victims' artlessness and defenselessness: The drone does not announce itself, flies quietly, and attacks from such a long distance that escape is hardly possible, not to speak of defense. It should, therefore, be considered whether the use of drones cannot be justified at all for this reason already. One argument that can be made is the deviousness of this method that uses lethal weapons like a "bolt from the blue", without giving those affected the chance to protect themselves on time. Meanwhile, in a declared war—which has been presupposed here up until this point—there is no such protected "state of trust"; because in such a situation one must expect an attack at any time. This corresponds with the idea that other somewhat surprising methods of attack are not considered damnable *per se*, e. g. the use of warplanes that fly below the

[55] Apart from the means of espionage, this valuation Kant gives should still receive wide acceptance even today.

[56] This might be seen differently if one follows Kant's opinion that the mere act of espionage must be prohibited in itself (regardless of acts of killing that may be connected); cf. Chapter "Kant's Suggestion" above.

enemy's radar or are able to avoid or prevent being identified by using appropriate means. Nor is it, as a rule, possible for those under attack to anticipate the exact spot an artillery shell or tank shell hits the ground.

The demand for a complete exclusion of military means that make use of the victims' artlessness and defenselessness may, therefore, be desirable in itself, although it seems it will likely prove to be as unrealistic as the general prohibition of espionage Kant believed to be damnable.[57] However, the matter is different when drones are used with deadly force outside a declared war. In such cases there is indeed an international "state of trust", that an attack by air does not have to be anticipated at any given time. Even if there may be good reasons not to (formally) declare the war concerned, the "state of trust" remains standing. As a consequence, the use of drones appears to be damnable in such cases, because the homicide victims' artlessness and defenselessness are unjustifiably made use of, regardless of whether the weapons' release was authorized by a human using a remote-control or "autonomously" by the drone itself.

9 The Distinction Between Combatants and Civilians

Another possible avenue, different form that to establish evaluation rules, is to raise the question whether or not it is ethically problematic that a human carries out the killing of others by using a drone. In this case, primarily questions of imputation are involved.

A crucial criterion insofar for judging warfare by means of remote-controlled or self-controlling weapon-systems could arguably be the ability to distinguish between combatants and the civilian population. It is common knowledge that International Humanitarian Law prohibits targeted attacks against the civilian population without any military advantage.[58] At the utmost, so-called collateral damages may be accepted to a certain extent. But this is not a matter of justification of collateral damages, but rather of the antecedent question whether or not remote-controlled or self-controlling weapon-systems are actually able to distinguish between combatants and the civilian population at all. For example, mines typically cannot do so, which is why there are good reasons to prohibit them by International Law, whereas there may be exceptions, if the manner they are laid in ensures, to a large extent, that they can only become dangerous for combatants, for example by mining naval ports, or it's access roads etc.

Now, admittedly, there is a thesis that, for example armed drones, contrary to mines, are indeed capable of differentiating between military targets and civilian

[57] Apart from espionage, regarding which a ban by International Law is not even being discussed at all, the "vile practices" Kant has named are mostly prohibited (e.g. by the dVStGB); cf. once again the references given in fn. 27.

[58] Cf. e. g. Arendt (2014), pp. 27–30.

targets.[59] Regarding remote-controlled drones, it is alleged that by use of the installed cameras, the drone operator, i.e. the person actually deploying the weapon-system, can clearly distinguish between military targets and civilian targets. Thus, by using weapons-grade drones it is perhaps ensured even better than in cases of immediate front-line duty by soldiers that civilian targets are not hit as far as possible. From a principle perspective, little can be objected to this. For indirectly taking notice of information on war-related events is most likely not prohibited *per se*, because it will often (for example, at night) make even more accurate information available than immediate human perception.

Going beyond this, it can be argued that persons immediately engaged in war-related events are under higher psychological pressure and thus tend to make wrong decisions under this pressure, while whoever operates the drone's weapon-systems from a distance can make more emotionless and therefore rational decisions, only limited by the rules of appropriate warfare. In doing so, the usage of armed drones could, on average, lead to less collateral damage or evident mis-determinations at the expense of civilians than the immediate decisions made by soldiers in the field.

Nevertheless, some doubts may remain as to whether or not the "camera perspective" can indeed produce this degree of certainty in differentiating between military and civilian targets that is asserted.[60] However, the argument that this actually is the case, cannot be rejected *a limine*. This is different, and in fact *principally* different, if so-called autonomous drones are used, i.e. weapon-systems that make the "decision", whether or not to fire the on-board weapon-system, independently ("autonomously") based at least partly on information they "ascertained" only themselves. In the course of this, it must be resolved whether such a "robot", that a drone can be seen as, is actually able to "decide" anything (in a reasonable sense of this word).

10 Problems of Imputation

In answering this question, a systematic approach that is generally used in Criminal Law can be helpful.[61] Criminal Law distinguishes between two levels of imputation when it imputes responsibility for a certain incriminated event.[62] On the one hand, imputation of an action, on the other hand, imputation of culpability. Simplistically paraphrased, the *1st level of imputation* deals with the question whether a certain result can be imputed to a person as a result of his action; if, for example the death of a person M was caused by the action of another person P. This can only be said

[59] Cf. Banaszewska (2014), Leveringhaus and Giacca (2014).

[60] Cf. Leveringhaus and Giacca (2014), p. 15.

[61] In more detail Joerden (2015) with further references.

[62] For additional questions of imputation in the context of the use of drones cf. Frau (2014b), pp. 238–243.

if P was *free* in his action. P is not free, if he had no alternative for action or no possibility of foreseeing the occurrence of the result (i.e. had neither current nor potential sense of acting). If this freedom of action is present, however, one can say that P has committed an act of killing.

In saying this, P's responsibility for M's death is still not yet determined. Because after criticizing his action as illegal, it is necessary to resolve whether this infraction of law can be imputed to the person P as *culpable* on the *2nd level of imputation*. This is only the case if P was also *free* regarding the *evaluation* of his act as illegal. He is not free, if he was placed in a relevant state of emergency or had no perception whatsoever regarding the wrongness of his behavior (i.e. had neither current nor potential sense of wrongdoing). Only if these requirements are met, the act of killing can be imputed to a person P as culpably caused by him.

Now one might suggest that a so-called autonomous drone can *act* more or less freely on the basis of the data it collected from the surroundings of war-related events, and that it has freedom of action in this sense. But even with an ever so high degree of (seeming) "autonomy" a drone will not be able to decide in a rule-governed way, whether or not the firing of a deadly weapon is permissible in accordance with International Humanitarian Law. This requires (in the Criminal Legal language) that the drone would be able to develop a sense of wrongdoing, i.e. the ability to subsume its action under the general rule of prohibition of killing civilian targets. This further requires that the drone actually has an idea of what a military target is, opposed to a civilian target. However, it fundamentally does *not* have such an idea, because this is not a cognitive but in the broadest sense a *normative* question of *evaluating* one target as military and another target as civilian that cannot be reduced to the coding in a binary computer language.[63]

Considering a drone's principal lack of competence of evaluation regarding the differentiation between military and civilian targets, the usage of such so-called autonomous drones with their own possibility to fire arms, must be rejected. Their usage would be equivalent to letting a random generator conduct the differentiation between military and civilian targets in the course of military conflicts. This, however, appears to be an apparent violation of International Humanitarian Law and of the ethics behind its rules. If one wants International Humanitarian Law to be observed in advance in every case in warfare, then there must at least be someone who is able to do so. But in the case of a so-called autonomous drone there is no one who is able to evaluate the case. Not the constructor of the drone, because he only can decide in general and not in the individual case. And this individual case is not foreseeable for him. So he cannot take care of complying with the norms of International Humanitarian Law. But also, the drone itself is not able of complying, because it has the already mentioned principal lack of competence to evaluate in accordance with the law.

---

[63]In the end, this is also the reason why judges can principally not be replaced by computers.

References

Arendt R (2014) Der Einsatz autonomer Waffensysteme im Lichte des Verhältnismäßigkeits- und des Unterscheidungsgrundsatzes. In: Frau R (ed) Drohnen und das Recht. Völker- und verfassungsrechtliche Fragen automatisierter und autonomer Kriegsführung. Mohr Siebeck, Tübingen, pp 19–34

Banaszewska D (2014) Kombattanten und Zivilisten weit weg vom Schlachtfeld. In: Frau R (ed) Drohnen und das Recht, Völker- und verfassungsrechtliche Fragen automatisierter und autonomer Kriegsführung. Mohr Siebeck, Tübingen, pp 59–81

Borrmann R (2014) Autonome unbemannte bewaffnete Luftsysteme im Lichte des Rechts des internationalen bewaffneten Konflikts. Anforderungen an das Konstruktionsdesign und Einsatzbeschränkungen. Duncker & Humblot, Berlin

Eser A, Sternberg-Lieben D (2014) In: Schönke A, Schröder H (eds) Strafgesetzbuch. Kommentar, 29th edn. C.-H. Beck, Munich

Frau R (ed) (2014a) Drohnen und das Recht. Völker- und verfassungsrechtliche Fragen automatisierter und autonomer Kriegsführung. Mohr Siebeck, Tübingen

Frau R (2014b) Völkerstrafrechtliche Aspekte automatisierter und autonomer Kriegführung. In: Frau R (ed) Drohnen und das Recht. Völker- und verfassungsrechtliche Fragen automatisierter und autonomer Kriegsführung. Mohr Siebeck, Tübingen, pp 235–250

Hruschka J (1987) Imputation. In: Albin E, Fletcher GP (eds) Rechtfertigung und Entschuldigung, Bd. 1, Freiburg i. Br., pp 121–174

Hruschka J (1991) Verhaltensregeln und Zurechnungsregeln. Rechtstheorie 22:449–460

Joerden JC (2005) Über ein vermeintliches Recht (des Staates) aus Nächstenliebe zu foltern. Annu Rev Law Ethics 13:495–525

Joerden JC (2007) Spuren der duplex-effectus-Lehre im aktuellen Strafrechtsdenken. In: Pawlik M, Zaczyk R (eds) Festschrift für Günther Jakobs zum 70. Geburtstag, Carl Heymanns, Cologne, pp 235–257

Joerden JC (2011) WikiLeaks, Kants "Princip der Publicität", Whistleblowing und "illegale Geheimnisse". Annu Rev Law Ethics 19:227–239

Joerden JC (2014) Die Differenz zwischen Mord und Totschlag im deutschen Strafrecht als Orientierungshilfe bei der Identifizierung von inakzeptablen Methoden der Kriegsführung, insbesondere im Hinblick auf den Einsatz von sog. Drohnen. In: Pływaczewski EW (ed) Current Problems of the Penal Law and Criminology/Aktuelle Probleme des Strafrechts und der Krimnologie. Wolters Kluwer, Warszawa, pp 65–81

Joerden JC (2015) Lob und Tadel. Relevante Regelarten und ihr Zusammenwirken. In: Arnold S, Lorenz S (eds) Gedächtnisschrift für Hannes Unberath. C.-H. Beck, Munich, pp 221–240

Kant I (1795) Towards Perpetual Peace. A Philosophical Sketch, translation into English based upon Jonathan Bennett. http://www.earlymoderntexts.com/assets/pdfs/kant1795_1.pdf and http://www.earlymoderntexts.com/assets/pdfs/kant1795_2.pdf

Kant I (1797) Über ein vermeintes Recht aus Menschenliebe zu lügen. Kants Werke, Akademie-Ausgabe, Band 8:425–430

Kubik A, Zimmermann T (2013) Mord und/oder Totschlag? – Eine kritische Analyse aktueller Reformvorschläge zur vorsätzlichen Tötung. Strafverteidiger 2013:582–589

Lackner K, Kühl K (2014) Strafgesetzbuch, Kommentar, 28th edn. C.H. Beck, Munich

Leveringhaus A, Giacca G (2014) Robo-Wars. The regulation of Robotic Weapons. Oxford Martin School, University of Oxford

Löffelmann M (2013) Rechtfestisung gezèlter Tötungen durch Kampfdrohnen? Juristische Rundschau 2013:496–513

Mahn-Gauseweg C (2014) Automated Warfare – Operationen unbemannter Systeme. Eine technische Systematisierung. In: Frau R (ed) Drohnen und das Recht. Völker- und verfassungsrechtliche Fragen automatisierter und autonomer Kriegsführung. Mohr Siebeck, Tübingen, pp 1–17

Merkel R (2012) Die "kollaterale" Tötung von Zivilisten im Krieg. JuristenZeitung 2012:1137–1145

Werle G, Jeßberger F (2014) Principles of International Criminal Law, 3rd edn. Oxford University Press, Oxford

Wolf G (1996) Befreiung des Strafrechts vom nationalsozialistischen Denken? Juristische Schulung 1996:189–195

Part III
Unmanned Systems in Particular at Sea and Land

Autonomy of Mobile Robots

Georg Heppner and Ruediger Dillmann

Abstract What do we actually mean when we talk about "autonomous mobile robots"? This chapter gives an introduction into the technical side of autonomy of mobile robots in the current state of the art and provides the relevant technical background for discussions about autonomy in mobile robot systems. A short history of robotic development explains the origins of robotics and its development from industrial machines into autonomous agents. The term "autonomous robot" is looked at in more detail by presenting the way of decision-making with different categories of robots and by introducing the general model of a rational agent for decision making. Additionally a short outlook on the process of understanding the environment is discussed, giving an overview of the individual steps from sensing up to interpretation of a scene. Selected examples of modern robots present the current state of the art and its limitations within this field. Overall, this introduction provides the required technical insight for non robotic experts to get an understanding of the term autonomous mobile robots and the implications for regulations concerning it.

1 Introduction

Most people today are familiar with robots of all sort as they are not only used for industrial production or in state of the art laboratories but are more and more becoming part of our daily lives. Better computers, smaller sensors and reduced production costs for parts that once would have required major resources have lead to affordable and versatile robots that are used in applications ranging from specialized medical equipment like the DaVinCi System, to household appliances like the vacuum robot Roomba or even toys such as the robot dog Aibo (Fig. 1).

While the prevalence of robotic systems increases, much of the expectations about them is still built up by movies where robots are generously used as a tool for every purpose. They are portrayed as intelligent, often almost human-like, entities that are just much faster, stronger and overall more precise than humans in any

G. Heppner (✉) • R. Dillmann
Interactive Diagnosis and Service Systems, FZI Research Center for Information Technology, Karlsruhe, Germany
e-mail: heppner@fzi.de; dillmann@fzi.de

W. Heintschel von Heinegg et al. (eds.), *Dehumanization of Warfare*,
DOI 10.1007/978-3-319-67266-3_5

Fig. 1 Robot dog Aibo which is sold as a toy

situation. It is therefore reasonable that there are also concerns about autonomous systems, especially if looked at in the context of military applications.

- But what really constitutes a modern "autonomous robot"?
- What does the term autonomous actually imply?
- What are the capabilities of autonomous robots and where are the boundaries?

This chapter will give a high level introduction to this vast research field[1] and look at some examples to illustrate the current state of the art in autonomous robotic systems. In Sect. 2 we will take a short look at the history of robotics and the development of them into mobile robots. Section 3 gives a definition of the term autonomy itself before Sect. 4 provides some more insight into how robots perceive their environment. Finally Sect. 5 will present selected robots and look at some of the decisions they make. Section 6 summarizes current capabilities of robotics systems, but also their limitations. Section 7 finishes with a short conclusion.

2 History of Robot Development

The dream of mechanical devices that do the work of humans is much older than one would expect. Some of the earliest mentions of mechanical devices that could fit the description of a robot date back to 400BC when the Greek philosopher Archytas invented a steam powered pigeon. One of the first production aiding robots was the "Wodden Ox"[2] (between 400BC and 230AD). The pure wooden wheelbarrow in the shape of an ox was (supposedly) equipped with legs that facilitated the energy from up and downward movements for thrust, allowing to carry loads of 200 kg for

[1]This introductory text is targeted at non robotic experts. Its purpose is to give a concise introduction and overview rather than to provide a complete and exhaustive discussion.

[2]Yu-Gang and Meng-Hui (2007).

up to 10 km per day. This very early example indicates the goal of such systems: Machines that can do work or movement on their own and help humans in tasks that would otherwise be tedious, hard or impossible.

The term "robot" stems from the slavic word "robota" which means "subordinate labor". It was coined around 1920 by Czech playwrighter Karel Ćapek in his play "Rossum's Universal Robots (R.U.R)".[3] While these robots where science fiction, the first real devices called robot made their appearance at around 1960 with advances of Computerized Numerical Control (CNC) and teleoperated master slave systems for nuclear waste handling. CNC machines consist of motors, encoders to detect the motor position and micro-controllers that can be used to repeatedly move axles to precise positions which are either preprogrammed or calculated based on analytic equations. One of the first robots in this category was the "Unimate", which was installed in a General Motors plant in 1961.[4] They were used for workpiece handling and spot-welding to great success, spawning many iterations with different forms and coined the term "industrial robot" which became an essential part of flexible manufacturing systems.

Industrial robots know next to nothing about their environment and reproduce the same movements over and over again with great precision. With the advancement of processors and better sensor technology the systems became more flexible. They could be reprogrammed or adapt to certain situations (for example, by estimating a workpiece position with a camera), but were still restricted to very closely defined operating parameters given by the human operator. The Robotic Industries Association defined the term industrial robot like this: "An 'industrial robot' is an automatically controlled, reprogrammable, multi purpose manipulator, programmable in three or more axes which may either be fixed in place or mobile for use in industrial automation applications".[5]

In 1980 the term robotics was defined as "intelligent connection between perception and action".[6] This definition emphasises on a control that is reactive to a given sensor input, rather than just executing predefined movements with small variations. Robots, by such a definition, already incorporate some form of low level autonomy. For example, the movement system of a mobile platform may decide which actuator to activate and at what speed to achieve a given goal, such as driving in a circle or a straight line. During execution it may also change these values based on sensor input to react to disturbances, wheel slippage and other unforeseen influences.

Around the year 1990 a demand for robots capable of tasks in hazardous environments, augmentation of humans or even services for everyday tasks led to the development of field and service robots. These field and service robots use the same basic principles as their industrial counterparts for movement, actuation or control but are fundamentally different in their application environment which has lead to

---

[3] Siciliano and Khatib (2008).

[4] Siciliano and Khatib (2008), p. 964.

[5] Kumar et al. (2008), p. 89.

[6] Siciliano and Khatib (2008).

very sophisticated approaches for many problems. One of the main differences is that these robots have to work in a "real world" environment, in contrast to industrial robots that operate in a well defined environment and without much deviation from the preplanned path. Simple "adjustments" to predefined actions are therefore no longer sufficient to achieve a certain goal, instead a solution has to be calculated for every new situation, resulting in a significantly changed behaviour of such systems. Additionally the actions a robot has to perform to reach a certain goal might also change, even during execution. With a rapidly changing environment to react to and complex mission goals, the autonomy of a system becomes one of the main requirements, often outweighing precision or repeatability.

Since then robotics has improved quite significantly, resulting in complex machines which are not limited to either industrial or very hazardous environments but appear in almost every area of our lives, if however not always directly visible as such. A modern cellphone contains enough computational power and sensors to build (together with some actuators) mobile robots that can sense their environment and react to it in realtime, while actuators have become so small that it is possible to build artificial bees or birds that almost look like their biological examples. One could argue that even our cars are teleoperated robots, as the input of a modern steering wheel is interpreted by sensors, evaluated by a computer and turned into movement of the axle by computerized actuators. Some manufacturers even provide first "autonomous" functions such as keeping a fixed distance, automatic parking in a tight spot or even the complete steering of the car on certain roads. With all these advances, the autonomy of such systems has become a major factor for their effectiveness but also a critical issue for law and regulations, which begs the question of what do we mean exactly when we talk about the term "autonomy".

3 Autonomy of Mobile Robots

A distinction should to be made between the term "autonomous" and "automatic" or "automation".

Industrial robots (Fig. 2 left) are a prime example for automation. Welding of car parts for example is almost completely handled in automated processes during which a multitude of actions are performed. Once the welding process is started the robot will follow a predefined path without any human intervention. During this process there might be several sub-steps such as switching the tool, rotating the workpiece or even reacting to sensory input, for example detecting if a new workpiece is present. No intervention of humans is needed, thus making the complete process "automatic". However, it is not considered to be "autonomous" as the actions that are taken by the robot are always the same and do not deviate from initially given instructions. In case one of the actions fails there is usually a form of recovery mechanism, but ultimately the whole automated process will stop if there are no predefined options left for a given situation.

Fig. 2 *Left*: Industrial robots in a car factory can perform fixed tasks with great speed and accuracy; *Right*: Automated guided vehicles which are used for logistic tasks in hospitals. *Left*: "BMW Leipzig MEDIA 050719 Download Karosseriebau max" by BMW Werk Leipzig http://bmw-werk-leipzig.de. Licensed under Creative Commons Attribution-ShareAlike 2.0 via Wikimedia Commons—https://commons.wikimedia.org/wiki/File:BMW_Leipzig_MEDIA_050719_Download_Karosseriebau_max.jpg. *Right*: "Automated Guided Vehicles" by MrGRA—own work. Licensed under Creative Commons Attribution-ShareAlike 4.0 via Wikimedia Commons—https://commons.wikimedia.org/wiki/File:Automated_Guided_Vehicles.jpg

Encyclopedia Britannia defines the term "autonomy" as: "(Greek, self and law), in general, freedom from external restraint, self-government".[7] The term is mostly used in conjunction with another adjective describing what part of a system is actually meant. Computational autonomy is one of the key requirements for mobile robots indicating nothing more than that the system is carrying its own processors and computing all actions on-board, energy autonomy denotes the fact that a system is carrying its own power source and is not dependent on outside power.[8] If used without further description the term "autonomy" is referring to *decisional autonomy*, meaning that an entity can decide what to do by itself. But still, this definition can vary upon the context, so it is useful to look at different areas for clarification. As traditional industrial robots (see Fig. 2 left) require little to no autonomy, we will focus on mobile systems.

3.1 Mobile Industrial Robots

Mobile robots can come in many forms depending on their application scenario but all require at least some form of autonomy to perform their tasks. We will use an automated guided vehicle (AGV) (Fig. 2 right) as example in this section. AGVs are mobile platforms that can transport cargo along predefined or self selected routes and can react to sensor input to perceive their environment. They usually interact with a centralised warehouse management system and possess the basic

[7] Various (1911).

[8] At least for a certain amount of time.

capabilities for mobility while executing a task, i.e. they possess energy autonomy, computational autonomy and so on.

Let us assume the following scenario: In a warehouse, five AGVs are operating in parallel to fetch items and bring them to a drop off point. Only one vehicle may enter the drop off point at any time and each order to the warehouse system will allocate a time slot for the vehicle at the drop off point. During execution of a transport job there are workers blocking the way of the vehicle resulting in the AGV to stop and subsequently arrive at the drop off point delayed. As another AGV might already be using the drop off point at that time the AGV (or the warehouse management system) has to make a decision about how to handle this situation. The AGV might change its route to make up the lost time or on the other hand delay its path by taking an even longer one to arrive at a later available timeslot. If the delivery is time critical, the warehouse management system might even decide to delay all other deliveries to clear the drop off point.

There are many similar situations that require decisions. Does the AGV wait if a passage is blocked or try an alternative route? Is the detected obstacle actually a human worker? And so on.

The system has to actively make an *autonomous* decision which depends on the environment, the current state of the robot and the overall goal that has to be achieved. But all decisions are happening in a closed environment, greatly reducing the level of required autonomy. While switching routes is an option, the AGV will never leave the predefined paths and can only choose between several options to handle a problem. In this example: *waiting*, *taking another route* or *asking to delay*. This is a classic scheduling problem that can be solved analytically or by simple rule sets. Most importantly, the sensor information (or more precise: their interpretations) are limited in their possible outcomes. Something in the way of the AGV is always an obstacle and the path can be detected by very precise sensors (for example within the floor of the warehouse), leaving little room for uncertain conditions.

3.2 *Mobile Field and Service Robots*

Field and service robots can also take many different forms and are used in many applications. The term service robots is usually associated with any robot that is actively assisting a human, for example by cleaning up the home or by helping with disabilities while field robots (like LAURON V,[9] see Fig. 3 right) are used in applications such as search and rescue after disasters or exploration of hazardous or remote areas. While the definitions can vary and often blend into one another, a key commonality is that these robots have to operate in an, at best, partially known environment that is highly dynamic.

[9]Roennau et al. (2014).

Fig. 3 Service robot HoLLiE (House of Living Labs Intelligent Escort) was build from commercially available parts and is mainly used for mobile manipulation research. LAURON V is a bio-inspired field robot that can be used to explore difficult areas, carry sensor payloads and use its legs for manipulation tasks. Field and service robots like this can be used in personal homes, public facilities, outer space or disaster sites. As such robots are mobile they have to work constantly changing environments, requiring a very high level of autonomy

We will look at the service robot HoLLiE[10] (see Fig. 3 left). This humanoid robot is designed as a service robot that can, amongst other things, assist with various household tasks like picking up the laundry and putting it into the washing machine. The task of picking up laundry requires a very high level of autonomy, as there is no predefined solution to follow. First, the robot has to find laundry on the floor by interpreting camera images whereby the floor, static objects and movable objects such as the laundry have to be discerned and precisely located in respect to the robot. Once the laundry is found, the robot has to pick it up with its hand which requires a combined movement of the platform, the arms and the body while factoring in every obstacle along the way. Such movement alone can be done in a multitude of ways to accomplish the same end result, for example by lowering the arm first and then driving towards the laundry or the other way round. If the environment is completely known by the robot, the selection of the right way to accomplish the task can be done by calculating the energetic cost of each action or the time the action would take. But the robot must also consider that there are obstacles behind other objects which can not yet be observed by cameras or other sensors. This results in a multitude of possible actions that not only vary in their known costs, such as time it requires to follow a specific path, but are also based on probabilities like

[10]Hermann et al. (2013).

the likelihood of other obstacles blocking the way. Finally, bringing the laundry to the washing machine requires skills such as navigation and obstacle avoidance in an environment that is constantly changing because furniture is moved, doors are closed or simply because humans are standing in the room. Accomplishing these tasks requires to make a multitude of decisions that can not be put into simple terms such as "if there is laundry, pick it up". In contrast to the industrial counterparts service robots must usually rely on their own sensors and can't use high precision equipment that is embedded in the environment. Additionally, the possible solutions aren't that closely restricted. The robot may move anywhere, resulting in unforeseen behaviours that can even be plain wrong. To understand how a robot can make such a decision we need to understand the idea of a rational agent.

3.3 Rational Agents

While there are different approaches to decision making and "intelligent" robots, we will focus on the approach to model robots as a rational agent. The book *Artificial Intelligence—A modern Approach* presents the term rational agent like this[11]:

> "A rational agent is one that acts so as to achieve the best outcome or, when there is uncertainty, the best expected outcome."

While this definition sounds simple enough, there are two distinctive parts that pose an enormous challenge for designing such a rational agent.

The best outcome needs to be translated into something that is measurable and decidable by the robot. The reason we see (advanced) robots do outright false things is often not the problem of the robot using the "wrong" action but rather that the definition of "good" was incomplete, inappropriate or simply incorrect. Also taken from the book Artificial Intelligence—A modern Approach, this can be clarified with a simple example: They propose a vacuum robot in the vacuum world (see Fig. 4). The world consists of two rooms which are connected to each other and are otherwise sealed. Each room can either be dirty or not. The Vacuum robot can move left, move right, do nothing and clean the floor. What would be the rational thing to do to keep the floors clean?

The initial answer is: Move left, clean, move right, clean, repeat. Such behaviour will most certainly keep the floors clean, but we might not see it as rational right away as the robot is constantly moving, even when the floor is completely clean. The problem in this case is the definition of the performance measure which only states that the floor must be clean. If the movement is penalized a better answer

[11] Russell and Norvig (2002), p. 32.

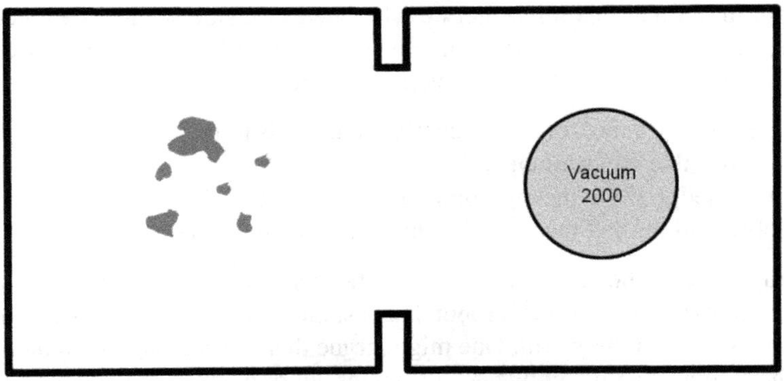

Fig. 4 Vacuum robot in the vacuum world. This example with greatly reduced complexity can be used to explain the rational agent approach

might be to wait a while after returning to one of the rooms. This problem becomes even harder when the resulting outcome is not directly measurable. When we think about the AGV example in the previous sections it is not easy to determine what effect waiting or choosing a shorter route would have on the overall performance of the warehouse. In fact, many companies put great effort into their warehouse logistics, trying to maximise the decisions of the autonomous system by analysing huge chunks of data to find the "best" cost function to optimise different, sometimes even conflicting, measures. A fast delivery is not necessarily a cheap one, while the cheapest short term solution might use up the most durability, costing more in the long run and so on. Of course the right action to take also depends on the actual means the robot has. With each additional possible action the robot can perform the robots behaviour changes.

The best expected outcome goes one step further, concerning especially the area of field and service robots and should be read with an emphasis on *expected*. If the world is not entirely known, the robot has to gather knowledge about it with its sensors or assume things based on the a priori knowledge it might have. Let us look again at the vacuum example. It is clear that the robot should only move to a room and clean it when it is actually dirty. If the robot does not know about the state of the room it can only assume. Based on the given fact that a room might be dirty, the robot could clean every room at a given interval. Probably the best solution would be to add a sensor to the robot that detects if either room is dirty. Then the cost function and resulting actions would be well defined: Move to the room that is dirty and clean it; If no room is dirty, do nothing. Unfortunately sensors do not work like this outside the vacuum world, always leaving some uncertainty in the measurements. If the robot used a camera to detect a dirty floor by searching for irregularities, the setup would most likely work well in a clean environment with industrial floor. But if one of the rooms has a carpet with a random pattern, the sensor might easily be fooled and would detect a dirty room, at least in some conditions. As the robot is not omniscient, it has to rely on what it *expects* to be the best action, and it is important to keep in mind that this might not be what we as outside spectator expect as the robot might conceive the circumstances in an entirely different way than we do.

Overall the behaviour of the rational agent depends on:

- The performance measure/cost function that needs to be maximised
- The actions that are available to it
- The knowledge about the environment
- The observations that the agent has made up to the decision

Finally, state of the art goes even a step further in defining the term Autonomy. All previous examples assume a robot that is static in his abilities, sensor evaluation and its modelling of the world. One might argue that, even though field and service robots have to deal with complicated decisions, all of them are only "automatic", as the developers will define the cost function, the sensor evaluation routines and the model that is used to understand the world. Even when there are large uncertainties, the overall behaviour is well defined and pre-selected just as it is with industrial robots. When following this approach, the only real autonomous systems are those that can learn new things. When we think of the vacuum robot, this could mean that the robot will "learn" what the adjective "dirty" actually means, and will not only clean the floor when there are irregularities but also when the color has changed significantly. Learning with robots is a huge field, ranging from quite simple approaches that basically try to find the right values for a specific function up to complex models that try to reason about the world from learned terms. The currently available vacuum robots for example will "learn" the layout of the area they are cleaning, decreasing their cleaning time in subsequent runs as they will not run into obstacles that often. While this topic is of course very promising for future developments, it is too complex to be adequately discussed in this text. More importantly, most of the robot systems that are used in any application today use only very limited forms of learning which doesn't give them much more autonomy than their non-learning counterparts.

In consequence, when we talk about autonomous mobile robots, what we usually refer to is:

> A field or service robot with computational, energy and movement autonomy that acts like a rational agent with different levels of sophistication concerning its sensors, performance measures, models of the world and its capabilities.

4 Understanding the Surrounding

In the previous section we have seen that sensing the environment is a crucial aspect for the autonomy of mobile robots. Therefore we want to take a short look at how robots actually perceive their surroundings.

Sensors come in all forms, sizes and levels of accuracy that need to be carefully selected to create a robust solution. Humans have the luxury of a very complex vision system that is perfectly balanced for our environment. Our eyes see in 3D and color at various distances, lighting conditions and even at very high speeds which is possible by a combination of highly integrated sensor cells (retina cells), fast adaptation systems (iris), clever techniques (saccade movements) and a highly complex filtering and interpretations system (brain). While technological counterparts have come a long way and even exceed human capabilities in some areas (for example in the infrared spectrum) they are still very specialised to specific applications and come with various shortcomings that make perception a challenging task. Lighting conditions are a prime factor for detection algorithms to fail because parts of a captured image are either over- or under-saturated, fast movements require a high update rate, which in turn limits the possible resolution and so on. It becomes even more difficult when the observation to perceive is hard to detect with current technology—some creativity is required. External forces are an example of something that is difficult to measure. Strain gauge sensors can be used to detect forces exerted on (mostly) rigid objects, but they are very fragile and sensitive. Capacitor measurements only work with some materials and most of the time the solutions that are available today aren't as flexible as human skin. All of these tradeoffs require a robot to be equipped with a whole set of sensors rather than just a single good one and a careful placement in order to minimize adverse effects between them. Having a large number of sensors is of course a possibility but it makes the following step much harder.

Sensor fusion is required to combine all the different sources into a coherent sensory result that accurately depicts what is happening around the robot. This includes prefiltering of signals, combining them into a single observation and also estimating how reliable a perception actually was. When we filter a sensor result, we might (deliberately) exclude certain observations. When measuring distances of a slow moving robot with a laser, it is common to use a low pass or median filter to exclude single measurements that are wrong due to reflections. While making the measurement more robust for most applications, we will get wrong measurements if there is a fast moving obstacle as it would simply be filtered out. While humans work with similar filters, nature had a considerable longer time to perfect this filtering. Additionally, the individual sources have to be constantly evaluated as to how reliable they actually are. The wheels of robots for example are typically used as an odometer. Just like in every car, we can measure the speed the wheels are moving and calculate a distance that was travelled to estimate the robot's position. But when the ground is slippery, the travelled distance might not coincide with the actual travelled distance because of the wheel slippage. In such a case, switching to a different measurement method is required, for example by evaluating the movement of the world with a camera. Finally all sensors are merged together to produce a single observation which can then be used by the robot to update its model of the environment or react directly to it.

Modelling the environment is required if the robot wants to formulate a plan or further evaluate the situation. While it is possible to react directly to sensor input, for example by backing away from a wall if it comes too close, most robots will have some form of model of themselves and the environment. These models can be at every level of complexity and can influence the robot behaviour as greatly as the performance measure. When we look at navigation, one of the simplest models would be to represent the robot as a circle (or any other geometric shape) and all of the surrounding area as a grid of either free or non free cells. This might be enough for a vacuum robot, but as it does not represent any heights, slopes or any other differentiation about the environment the robot will probably not reach its full potential. A cardboard box, which might be easily pushable by the robot, is interpreted in the same way as a granite wall, making it an obstacle the robot will avoid. Such a problem can often be seen at robotic challenges where robots stop along the way because their laserscanners detect high grass as an obstacle which the robot decides it can not pass.

Interpretation is the final step of the perception chain that builds up on the modelling and adds some more semantics to the problem. This part is especially important when we talk about "intelligent" robots that are supposed to act in the right way given a certain situation, for example self driving cars. While modelling of the sensor information can separate different entities and give a single snapshot of a scene, we need an interpretation to make sense of what is actually happening. To interpret what is actually happening we set the detected entities into relation and track them over time to extract patterns. A car can be disregarded as an obstacle when it is interpreted as "moving away" or "on the other lane", while the car directly in front of the robot is of course a relevant obstacle and also an indicator for the robot's possible speed. For a short amount of time, such interpretation even allows to predict future situations which is important for actions like entering moving traffic or avoiding other cars at higher speeds.

Advances in computer efficiency, available sensor techniques and algorithms have made it possible to detect a wide range of situations, but it is important to realise that every decision a robot makes is based on this chain of sensor input, filtering, fusion, modelling and interpretation. A process that can be difficult at every step and lead to serious misconceptions.

5 Examples

With the knowledge about what we can expect from autonomous robots, we want to look at some examples to understand how far the actual development has come.

Fig. 5 The bio-inspired snake like robot KAIRO III can reach difficult areas and can be reconfigured to match the application requirements

5.1 *KAIRO III*

The KArlsruhe Inspection RObot (KAIRO)[12] (Fig. 5) of the FZI is a snake-like robot inspired by biology, constructed for inspection tasks in difficult to reach spaces such as pipes or collapsed houses. Its body consists of multiple segments, each connected to the next by three harmonic drive gearboxes. Each segment is actuated by two powerful electro motors, giving the robot enough strength to pull light persons. One of the segments contains the main computer of the robot, while others can house additional sensors or payloads for a given mission. The first segment holds a rotating laserscanner which is used to sense the environment in 3D and construct a 2.5 D[13] map out of it (the so called Plexmap[14]).

A speciality of the robot is its modularity with the ability to reconfigure itself. By adding or removing segments the robot can be varied in size and carried payloads. It can detect its configuration on its own and therefore decide on the best action to take based on its actual capabilities. This can be illustrated with an autonomous navigation experiment (see Fig. 6). We ordered KAIRO to drive some meters forward without any further constraint or information. The robot then used its laserscanner to create a 2.5 D map of the environment which revealed that the robot was placed directly in front of stairs and next to a ramp. Based on the information about how many segments are currently in use the robot calculated that it is able to overcome the steps and therefore planned a path over them. In the case of fewer segments the robot would have concluded that it is not able to reach the required height by lifting its segments and would have taken the longer route over the ramp. After the decision has been made, a path for the first module is planned

---

[12] Pfotzer et al. (2015).

[13] 2.5 D denotes maps that are basically two dimensional but contain information such as inclination or multiple height levels.

[14] Oberlander et al. (2014).

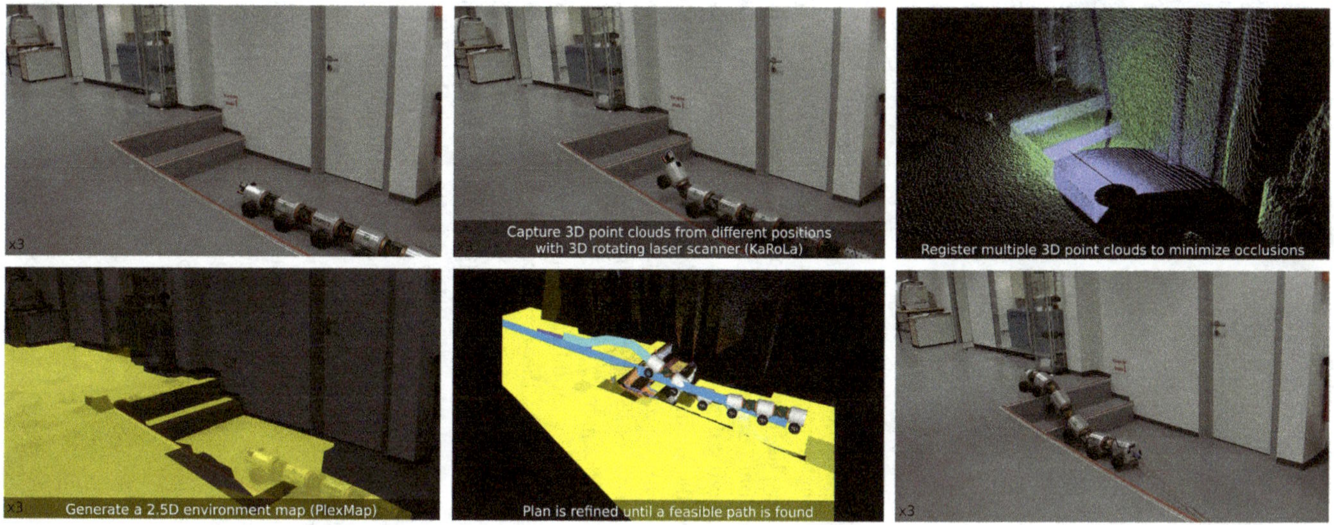

Fig. 6 KAIRO III overcoming stairs, a decision made by evaluating the environment and its current capabilities. A high precision laserscanner is used to create a dense 3D pointcloud of the environment. This data is then processed into a 2.5 D representation to allow easy path planning. Based on the number of segments the robot is equipped with a route is selected and a trajectory calculated. The trajectory is then transformed into motor commands to move the segments along the path

and corresponding motor actions for all segments are calculated to smoothly follow the first one.

5.1.1 Lessons Learned

The first thing to learn from this example is the environment sensing and resulting representation. The way this robot perceives its environment is quite common for many robot systems that operate in a 3D environment. Complex 3D environments are first sensed with appropriate hardware, in this case a laserscanner. While the resulting data is high in resolution, this technique requires a static environment, i.e. no people moving around the robot. It is possible to move the laserscanner faster to overcome this limitation at the cost of reduced accuracy and higher computational effort. Another option would be to use a stereocamera system, but this might have problems with the smooth industrial surface that is nearly uniform in texture. The resulting map contains inclinations and heights of the individual cells and is made specifically for ground robots and areas with varying inclinations, like the Moon or Mars, the core concept however is similar in other robots. A map like this enables the robot to plan movement paths, aid localisation and provides feedback to operators that can assess the environmental situation around the robot.

We can also clearly see how the available capabilities and the perceived environment is influencing the robot. In case of a shorter robot, the route would have been planned alongside the ramp, but only if the ramp was actually scanned correctly. KAIRO lifts its first segment to acquire a more complete picture of the environment. If it had taken the same scan from the ground level it would probably have seen no valid route whatsoever, resulting in either no movement or the execution of fallback strategies.

5.2 *CoCar*

The Cognitive Car (CoCar),[15] also from FZI, is a modified regular car that is used as a development, test and integration platform for driver assistance functions ranging from drive style estimation up to fully autonomous driving without user intervention. Based on a series production Q5, the car is outfitted with cameras in every direction, various stereo-cameras, time of flight depth sensors, radar, multiple lasers, odometers, differential GPS and inertia sensors (Fig. 7). The trunk is taken up by computer systems and electronics to support all these sensors and provide enough computational power to calculate complex algorithms on-board.

Many different aspects of autonomy are present in this robot as it can detect and classify lanes, intersections, other vehicles, pedestrians, traffic lights, the driver's

[15]Kohlhaas et al. (2013).

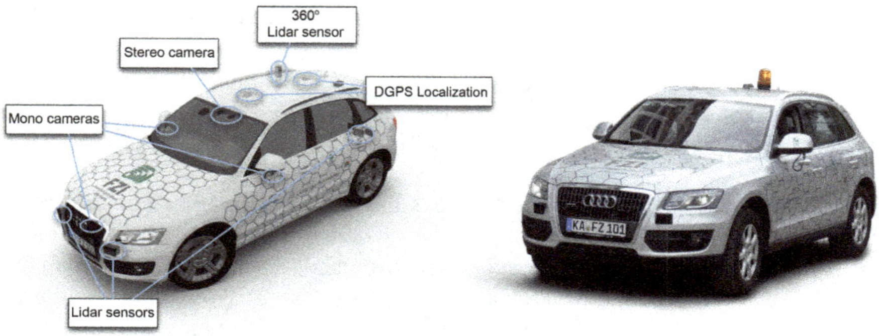

Fig. 7 The CoCar (Cognitive Car) by the FZI fuses many different sensor sources to get a clear understanding of its environment

gaze direction and many additional signals to create a complete representation of the surrounding environment. We want to focus on the extraction of meaningful information for the robot with the example of pedestrians crossing the street in front of the vehicle.

In the experiment (see Fig. 8),[16] pedestrians walk over a crossing in front of the car which has to stop and wait for them to clear the street. For humans, this task sounds easy enough, but for an autonomous system this entails a complex software pipeline to interpret the actual situation.

1. The area in front of the vehicle is monitored with a PMD (Photonic Mixing Device) depth camera and segmented by unseeded region growing
2. The selected areas are transformed back into a camera image for visualisation of objects and as selection for the classification step
3. Using Haar-Wavelet Ada Boost cascades the objects within the region are classified (in this case as pedestrians)
4. Movement of the classified objects is tracked with a Kalman filter
5. Finally, the attribute "blocking" can be assigned to the tracked objects by evaluating their type and spatial relation to the vehicle

By extracting the blocking attribute for the pedestrians the car can make the decision if it needs to stop or can continue driving. Just looking at the sensor information itself might not be enough in this case as a pedestrian might be detected waiting at the crossing without the intention of crossing it. When no tracking and further evaluation is done, this could lead to a full stop when not necessary which is undesirable for autonomous cars.

[16] Schamm and Zöllner (2011).

Fig. 8 Steps of the pipeline to get an understanding of the environment. *From top left to bottom right*: PMD sensordata is used to detect observations by unseeded region growing; Sensordata is transformed and fused with the RGB camera data; Segmented objects are shown within the RGB image; Classification of regions with Haar-Wavelet Ada Boost cascades; Tracking observations with a Kalman filter; Interpretation of the data to infer the attribute “blocking”

5.2.1 Lessons Learned

We can see from this example that a complex pipeline of filtering and detection is required to make sense of a scene. While the individual techniques may differ for other robots, the principle of understanding the environment will work the same. Without going into the details of the algorithms it should be obvious that the definition of the models within this example and the sensor performance are crucial for the resulting behaviour. The region growing needs to have parameters like minimal size boundary, the classification requires samples of pedestrians and might classify strange looking examples as something entirely different and dirt on the sensor might give a permanent observation in front of the car and so on.

On the other hand, we can see that even in a complex environment, modern robots are capable of understanding a scene and act accordingly. Only a single attribute is extracted in this example, which might become considerably harder when there are multiple or just more complex attributes such as "waiting" or when we want to determine if the pedestrian has actually seen the robot or not. Prediction is ultimately the next step in such a pipeline, giving the robot an idea wherever the pedestrian is actually going to cross the intersection or just waiting. Because humans can act very spontaneous, modelling such predictions is even harder and an active research topic.

6 Limits and Possibilities: Capabilities of Modern Robots

The two examples are of course just a very small subset of modern robots but they illustrate some core capabilities that we can find in most complex robots, mainly the pipeline of sensing the environment, understanding it and planning an action (in this case movement) with respect to it.

Limits of today's robots are given by two main factors:

The currently available technology, which limits miniaturisation of actuators and sensor technology resulting in the need for clever solutions that are often fragile and very specific to a certain scenario. Robots are by far not as strong as movies like to claim. As the power to weight ratio of nature has not been reached and speed is an adverse property to power, they are neither as fast nor as powerful as depicted. They are especially not as robust, as the technologies used today are highly integrated and often not hardened to mechanical stress. At least, they are not all of these things at once, which requires the specialisation to one of the attributes or a specific scenario. This is especially true for sensors which are constructed for a certain application scenario and it requires much effort to get a clear representation of the actual world, which will always be some abstraction of it.

Available algorithms, especially for understanding the environment and calculating the best possible action for it are highly complex and still a good bit away from being robust in all situations. While it is possible to understand situations as complex as traffic in a city, a simple misclassification of an object might result in a fatal conclusion. Many actions require not only an accurate understanding of what is happening at the moment, but what will happen in the future which is an even harder challenge. Therefore, modern robots can usually not enter crowds of people, react to situations they are not prepared for (at least not very well) and have difficulties selecting the right approach to a situation in advance. Additionally, robots may calculate some of the outcomes of their actions, but ultimately they can not predict what will happen like a human does, as they lack not only the computational power, but also the intuitive understanding of the actions' consequences that we posses. For example, a robot that throws down a glass will realize it is not available anymore, but not that driving over the shards might damage its wheels, unless it was explicitly stated in advance.

Another limiting factor is, that it is hard to learn something new. Robots are therefore still restricted to the possibilities the developers have integrated and can not learn on their own.[17] This is a problem, as it is almost impossible to imagine each and every situation in advance. Most learning is restricted to model identification, i.e. to learn the weight of the robot's arm by moving it around or the influence of rotor downwash in flying robots. If a scenario is not known, the robot might therefore simply fail, or decide in an undesirable way as it will not take everything into account.

Possibilities of robotics on the other hand are manifold for the right applications. When the requirements of detecting a situation correctly are ensured, the robot will do it safely over and over again. No sleep deprivation or simple distraction will influence the robot in its decisions as they are based on hard functions that take the available data into account. Additionally, the calculation of such decisions can be done instantaneous and by taking much more data into account than a human usually would, which is why driving assistance functions for cars have gained much popularity. Especially when we combine equipment that can fulfill tasks humans can't do, such as night vision cameras or heat sensors, robots can be used to monitor events that would otherwise be hard to spot.

When built for a specific task, robots are also quite capable in their actuation and movement skills. Walking robots that will overcome almost all obstacles as long as their fuel lasts are being discussed for resupply missions and could potentially search for injured people after a disaster indefinitely and without rest. Not having to breathe or the ability to withstand environment conditions such as great heat or cold, robots also excel in areas that are simply too dangerous for humans. Industrial robots can lift heavy objects and aid in construction tasks with great precision. With smaller actuators there are now robotic hands in human size capable of working with objects made for humans, such as power tools or control panels enabling a very wide range of applications.

The application scenarios are endless, but only, if it is a well defined one and the robot is especially built to match it.

---

[17]There are of course projects that focus on this issue, however, they are currently researched and not yet found in many robots that are used in real world applications.

7 Conclusions

In this chapter, we have taken a look at what the term "autonomy of mobile robots" actually entails. We first looked at the historic development of robots from simple machines with repeating tasks into versatile "intelligent" mobile robots that have to make difficult decisions in a wide range of environments. To explain how robots come to a decision and therefore act autonomously, we have used the rational agent approach which showed that the information a robot is having, as well as its performance measure play a significant role in the way a robot acts. We have further looked at the complex computation pipeline that a robot has to perform to come to a representation of its environment in order to understand what their decisions are actually based on. With this context we have taken a look at two state of the art robots in different domains, showing some key components and current techniques.

Robots can move in various ways, control complex kinematic chains (arms) and travel at high speeds. They are capable of understanding complex situations within a well defined subset of the world and can make decisions without active intervention. However, they are still restricted to special application areas, as their sensors might produce false results and because the knowledge we implement in them to make decisions might not cover all situations.

When talking about autonomy of robots, the human factor should not be left out of the discussion. Even if all the sensors work perfectly (which is unlikely) and every algorithm produces the perfect result (which is, given the uncertainties, almost impossible) we would end up with a decision that is based on a formal definition that was given in advance. Robots will not question this definition, which in turn begs the question: How well can we define a cost metric so that the outcome is what we actually want? While it seems to be easy to formalise decisions or goals, even simple examples like picking up laundry has many pitfalls. How do we exclude already washed laundry that is lying on the couch? Defining clear and decidable rules for these decisions is hard and often leads to unwanted results. The more critical the application gets, the more sure we should be about these definitions, which is why in safety critical applications the human will always have to make the final decision.

Finally, when we talk about robots, we must not forget that they are technical systems. Like every computer, they can be manipulated and altered to change their behaviour. This can be as simple as destroying a sensor and go as far as to hack the system and alter its performance measures. If we talk about decisions and regulations regarding those decisions, we must also make sure that the metrics we outfitted the systems with are itself protected from any tampering.

The term autonomy is a complicated one and even within the group of robot researches often disputed. This chapter's aim was not to define this term with absolute certainty but to illustrate some key points that need to be considered when talking about the legal implications of using such systems for difficult tasks:

- The behaviour of a robot is neither something completely predefined like the motion of an industrial arm nor completely surprising as it would be the case

with learning systems but is defined by a complex chain of sensing, coupled with various models and performance functions
- With complex state of the art systems, there is no absolute guarantee to control the outcome for every action, especially as there are still many challenges in sensor technologies and uncertainties within each step
- The way we define models and performance functions directly influences what a robot will do and need to be well defined to reach a desired behaviour.

Robots are a great tool for many applications. In certain areas they can perform incredible tasks and often they can do it autonomously without any intervention. But we have to keep in mind that this is not universally true. When we discuss laws and regulations about such autonomy, it is important to make clear what sort of autonomy or which part of it is actually addressed. It is important to understand that solutions are based on metrics and we need to define the rules in such a way that developers can actually put these metrics into the software and even more important, we need to make sure that these metrics actually depict what we want the robot to do. Terms like "certain" or "sure" need to be quantified with a number of tests and values like the rate of false positives in order to provide meaningful restrictions. And finally, we should remember: When we humans with all our knowledge and cognitive abilities make decisions, we still make mistakes quite frequently, why should a robot, without all these abilities and far less knowledge, perform any better?

References

Hermann A, Sun J, Xue Z, Ruehl S, Oberlaender J, Roennau A, Zoellner J, Dillmann R (2013) Hardware and software architecture of the bimanual mobile manipulation robot HoLLiE and its actuated upper body. In: 2013 IEEE/ASME international conference on advanced intelligent mechatronics (AIM), pp 286–292. https://doi.org/10.1109/AIM.2013.6584106

Kohlhaas R, Schamm T, Lenk D, Zöllner J (2013) Towards driving autonomously: autonomous cruise control in urban environments. In: Intelligent vehicles symposium (IV) (2013). IEEE, New York, pp 116–121

Kumar V, Bekey GA, Ambrose R, Lavery D, Sanderson D, Yuh J, Wilcox B (2008) Robotics: state of the art and future challenges. Imperial College Press, London

Oberlander J, Klemm S, Heppner G, Roennau A, Dillmann R (2014) A multi-resolution 3-d environment model for autonomous planetary exploration. In: 2014 IEEE international conference on automation science and engineering (CASE). IEEE, New York, pp 229–235

Pfotzer L, Staehler M, Hermann A, Roennau A, Dillmann R (2015) Kairo 3: moving over stairs & unknown obstacles with reconfigurable snake-like robots. In: 2015 European Conference on Mobile Robots (ECMR), pp 1–6. https://doi.org/10.1109/ECMR.2015.7324209

Roennau A, Heppner G, Nowicki M, Dillmann R (2014) LAURON V: a versatile six-legged walking robot with advanced maneuverability. In: 2014 IEEE/ASME international conference on advanced intelligent mechatronics (AIM), pp 82–87. https://doi.org/10.1109/AIM.2014.6878051

Russell SJ, Norvig P (2002) Artificial intelligence: a modern approach, 2nd edn. Pearson Education, Harlow

Schamm T, Zöllner J (2011) A model-based approach to probabilistic situation assessment for driver assistance systems. In: 14th international IEEE conference on intelligent transportation systems (ITSC)

Siciliano B, Khatib O (eds) (2008) Springer handbook of robotics. Springer, Berlin

Various (1911) The encyclopaedia britannica, vol. 2, 11th edn., Part 1. Encyclopedia Britannica, Chicago

Yu-Gang C, Meng-Hui H (2007) Searching ancient inventions from modern techniques-a research of walking horses with 8-link type leg mechanisms. In: Proceedings of twelfth world congress in mechanism and machine science

Feasible Precautions in Attack and Autonomous Weapons

Jeffrey S. Thurnher

Abstract The future of warfare will undoubtedly include autonomous systems capable of making complex decisions without human operator involvement. These systems will react blindingly fast, possess exceptional precision, and operate reliably and consistently without human supervision. While the promise of autonomy seems almost boundless, questions remain about the lawfulness of allowing such systems to select and lethally engage targets on their own. One of the consistent issues raised is whether nations wishing to employ autonomous weapon systems on the battlefield could do so in a manner that would comply with the fundamental principles of the law of armed conflict. This chapter addresses the crux of that concern by examining the legal impacts these future systems might have on targeting during international armed conflicts. In particular, the chapter focuses on the requirement of nations to take feasible precautions in attack and seeks to determine whether a nation employing autonomous weapons on a battlefield might be able to fully comply. First, this chapter strives to define what is meant by autonomous weapon systems. Second, the chapter will examine the technological advances and operational benefits which portend these weapon systems may become a reality in the future. The emphasis will next shift to the unique challenges the systems present to the requirement to take feasible precautions in attack. Lastly, the author concludes that while questions regarding precaution in attack requirements raise valid concerns, the use of autonomous weapons under many circumstances will likely be deemed lawful.

J. Thurnher, International and Operational Law Department, United States Army Office of the Judge Advocate General, United States Army, Judge Advocate. The views expressed are those of the author and should not be understood as necessarily representing those of the United States Army, the United States Department of Defense, or any other government entity.

J.S. Thurnher (✉)
International and Operational Law Department, United States Army Office of the Judge Advocate General, United States Army, Judge Advocate, Washington, DC, USA
e-mail: jeffrey.s.thurnher.mil@mail.mil

W. Heintschel von Heinegg et al. (eds.), *Dehumanization of Warfare*,
DOI 10.1007/978-3-319-67266-3_6

1 Introduction

Autonomy is becoming commonplace on the battlefield, with systems increasingly capable of performing actions on their own without the direct input of a human operator. However, it is the potential advent of autonomous lethal targeting capabilities that is raising the spectre of opposition and presents the biggest legal challenges under international law. Although fully autonomous weapons have not yet even been designed, a robust opposition to them has already formed. A coalition of nongovernmental organisations and human rights advocacy groups are campaigning for the adoption of an international treaty to pre-emptively ban the development and use of autonomous weapon systems, or systems which are capable of selecting and engaging targets on their own.[1] This coalition fears these weapon systems will be unable to fully comply with international law and would endanger civilians in a conflict area.[2] Similarly, a United Nations (UN) Special Rapporteur expressed concerns about autonomous weapons and proposed a moratorium on all related testing and development in an April 2013 report to the UN Human Rights Council.[3] Representatives of the UN Convention on Certain Conventional Weapons have begun discussing autonomous weapon systems as part of a series of experts meetings.[4]

Foremost among the legal concerns raised is the question of whether a nation seeking to deploy autonomous weapons into an armed conflict would be able to ensure that it had indeed taken all feasible precautions in attack. This tenet of the law of armed conflict obliges combatants to operate with a view toward minimizing harm to civilian populations. While no nation has yet declared an interest in developing fully autonomous weapon systems,[5] these positions will likely change as the technology becomes more sophisticated and reliable and as the operational benefits of autonomy continue to emerge. Nations wishing to deploy autonomous weapons in the future must resolve the complex challenges surrounding the obligation to take all feasible precautions in attack. This chapter will examine the

[1]More than forty nongovernmental organisations have formed the *Campaign to Stop Killer Robots*, an umbrella organisation dedicated to seeking a comprehensive and pre-emptive ban on the development, production, and use of autonomous weapons. Campaign to Stop Killer Robots (2013). http://www.stopkillerrobots.org/. Accessed 13 December 2016.

[2]Human Rights Watch is one of the founding organisations of the coalition. For a full description of their reservations and criticism of autonomous weapon systems, see Human Rights Watch (2012), p. 1.

[3]United Nations A/HRC/23/47, p. 21.

[4]Convention on Conventional Weapons CCW/MSP/2013/CRP.1, p. 4.

[5]In fact, nations such as the United States and United Kingdom have declared they are not pursuing such weapons other than human supervised ones. House of Lords Debate 26 March 2013 (The UK Ministry of Defense 'currently has no intention of developing [weapon] systems that operate without human intervention'.); United States Department of Defense (2012a), p. 3 (The United States has no 'plans to develop lethal autonomous weapon systems other than human-supervised systems for the purposes of local defense of manned vehicles or installations'.).

particular relevant requirements and explore how the unique aspects of autonomy may affect the legal analysis.

To address these complicated matters, this chapter is organised into the following subparts. Section 2 defines the levels of autonomy that future weapons will possess. Section 3 examines two key factors driving nations to increase their reliance on autonomy in weapon systems. The first component is the recent, remarkable technological progress demonstrated in artificial intelligence and machine learning. The second driver of this trend is the immense operational advantages expected to be gained through greater autonomy. Section 4 details the composite elements of the obligation to take all feasible precautions in attack with a focus on how autonomy may affect compliance with each. Finally, Sect. 5 concludes with a discussion of why nations desiring to deploy autonomous weapons onto battlefields will likely be able to comply with legal requirements to take all feasible precautions in attack in many imaginable circumstances and why a ban is unwise and unnecessary.

2 Levels of Autonomy

A useful starting point for any examination of autonomous weapons is to define the term autonomy and to distinguish the various levels of autonomy that future weapon systems might employ. When discussing technology, the term autonomy describes the ability of a machine or computer system to perform certain functions without human assistance or direct input.[6] In the context of autonomous weapons, the crucial functions that a weapon system might be expected to perform independent of human operator involvement include the targeting actions of finding, fixing, and firing on an objective. The United States (US) Department of Defense (DoD) devised a definition for autonomous weapon systems that has become widely accepted and referenced. It defines an autonomous weapon system as 'a weapon system that, once activated, can select and engage targets without further intervention by a human operator'.[7]

The DoD has further attempted to stratify its systems based upon, in essence, three levels of autonomy that the systems may possess. First, its lowest level of autonomy refers to 'semi-autonomous' systems or systems which can only engage targets that have been specifically selected by a human operator. This lowest level of autonomy includes weapons such as 'fire and forget' missiles, where the operator selects and fires the weapon at a target but then the weapon autonomously hones in on and strikes the target.[8] The second level of autonomy under the DoD approach equates to 'human-supervised autonomous systems'. With human-supervised systems, a human operator retains the ability to override the system at any point but

[6]Krishnan (2009), p. 45.

[7]United States Department of Defense (2012c), pp. 13–14.

[8]United States Department of Defense (2012c), pp. 13–14.

the system can otherwise act independently to identify and strike a target. The third and final level of autonomy applies to fully autonomous systems that can select and engage targets without further human operator involvement. Some groups opposed to autonomous weapons, such as Human Rights Watch, seemingly identify three similar levels of autonomy, but they prefer to use the terms, 'human-in-the-loop', 'human-on-the-loop', and 'human-out-of-the-loop' respectively.[9]

A slightly more nuanced approach to the various levels of autonomy is proffered by Dr. Armin Krishnan in his book *Killer Robots*. Dr. Krishnan describes four levels as part of a spectrum or a sliding-scale of autonomy. In general, the more independently a system can perform its tasks, then the greater the level of autonomy it possesses.[10] His starting point is with 'tele-operation', where a system is completely controlled by a human operator via a remote control and has no autonomy. Next, he describes 'pre-programmed' autonomous systems that are specifically programmed to conduct precise and non-varying missions. Then, Dr. Krishnan details an additional level of autonomy known as 'supervised autonomy', whereby a system autonomously controls the majority of basic functions but not the most complex ones, such as those associated with targeting decisions. Those complex decisions would instead be made by human controllers. Dr. Krishnan's final level of autonomy is known as 'complete autonomy'. Completely autonomous systems would theoretically receive human input on overall objectives but could otherwise 'find solutions and handle' many complex targeting problems.

In general, unless otherwise specified, whenever this chapter refers to autonomous weapons it is intended to indicate a weapon system that falls under the fully autonomous weapon system level of autonomy as described by the DoD. It also equates to Dr. Krishnan's complete autonomy level of autonomy with one exception. Dr. Krishnan seems to imply that to reach this level of autonomy an autonomous system must be capable of operating 'without the need for any human input'.[11] The discussion of autonomous weapons in this chapter, however, does not anticipate that autonomous systems will operate devoid of human input in the foreseeable future.[12] Instead, a fully autonomous system will likely never be completely free of human input.[13] At a minimum, humans will decide under what parameters and frameworks it will operate and when and where it will be deployed.

[9]Human Rights Watch (2012), p. 2.

[10]Krishnan (2009), p. 43.

[11]Krishnan (2009), p. 44.

[12]It is conceivable that advances in artificial intelligence technology in the future may allow systems to possess human-like reasoning. However, it is far from certain that the technology will successfully develop in such a manner, and even Dr. Krishnan contends that any such advances would be unlikely to materialize until well beyond the year 2030. Krishnan (2009), p. 44.

[13]Schmitt (2013a), p. 4.

3 Drivers of Greater Levels of Autonomy in Weapon Systems

Despite the current absence of fully autonomous weapons in any nations' arsenal of weapons, numerous experts predict that the use of such systems in armed conflicts will become standard within a period of 20 years.[14] Many States are actively exploring ways to increase the use of autonomous features into their military systems. Two major factors help explain this growing interest in autonomy. The first revolves around recent substantial leaps in technology and science that are making the possibility of these futuristic systems more certain. The second involves shifting operational realities that will likely drive nations to develop autonomous targeting capabilities.

3.1 Advances in Autonomous Technology

Throughout history, scientific advances have led to remarkable improvements in weaponry and military tactics. The current pace of technological developments, however, has been unprecedented. The recent advancements in computer science and artificial intelligence have led some experts to conclude that it is technologically feasible today to design a fully autonomous weapon system.[15] Computer software innovations are at the heart of these successes, as scientists have found new ways of harnessing the increased computational power available. By converging complex 'high-level planning' computer algorithms,[16] scientists are enabling computerized systems to increasingly make independent decisions and actions.[17] Ground-breaking software is progressively incorporating machine learning and other learning models into its designs, allowing the systems to innovate themselves and improve upon prior performance.[18] Tremendous strides have also been made in researching how to make human-system interaction more efficient and fluid. Many nations' militaries are eager to fully employ these advances.

Automatic and autonomous features have long been a part of the weapons of war. Decades ago, navies around the world began using naval mines capable of responding automatically to 'acoustic, electromagnetic or pressure' signatures.[19]

---

[14]Singer (2009), p. 128.

[15]For example, the former chief scientist for the United States Air Force postulates that technology currently exists to facilitate 'fully autonomous military strikes'; Dahm (2012), p. 11.

[16]Guarino (2013).

[17]Poitras (2012).

[18]Guarino (2013). For a more general overview of machine learning capabilities and possibilities, see Russell and Norvig (2010), ch. 18. For a discussion about how computer systems are learning, in approaches similar to how humans learn by examples, see Public Broadcasting Service (2011).

[19]von Heinegg (2011), p. 184 (asserting that such mines are 'quite common and legally uncontested').

Since the mid-1970s, the United States Navy and others have placed self-defence close-in weapon systems (CIWS) aboard its warships. These systems can detect the signatures of incoming missile and rocket attacks and automatically respond with lethal force to defeat them. Similarly, for many years, air defence weapon systems, such as the United States Army's Patriot Missile system and the United States Navy's Aegis combat system, have had the capacity to automatically defeat incoming artillery or missile attacks on ground and maritime forces.

Recent computing advances are making possible even greater leaps towards full autonomy, and newly designed military systems are showcasing this potential. For example, the latest developments in autonomous flight have been remarkable. The United States has demonstrated the ability to have a combat aircraft autonomously take off and land from an aircraft carrier. The United States Navy aircraft, known as the X-47B, successfully performed these difficult manoeuvres without the help of a human controller during testing in 2013. The X-47B then became the first craft to perform aerial refuelling autonomously during testing in 2015.[20] For the past few years, the United States Marines have successfully used 'K-MAX' variant helicopters to autonomously fly and deliver supplies to forward operating bases across Afghanistan. The Taranis attack aircraft, a British Royal Air Force-led project, is expected to be capable of supersonic autonomous flight within the next several years. Maritime systems similarly show tremendous promise. Several nations have successfully developed underwater systems capable of autonomously adjusting themselves to maintain their position in the water for extended periods of time. These achievements have led the United States to begin exploring designs for a vessel capable of autonomously hunting enemy submarines for months at a time.[21] Even many land systems, such as numerous sentry robots that South Korea has deployed along the Demilitarized Zone (DMZ), possess advanced autonomous features.

Perhaps the most dramatic autonomous advancements, however, have been made in the cyber domain. Many cyber programs are now being designed to autonomously identify vulnerabilities in a target system and determine on their own how best to infiltrate the system.[22] Programs that will employ and 'react to countermeasures' autonomously are also being actively pursued.[23] Some commentators contend that autonomous cyber weapons not only exist today, but also have already been successfully employed as a means of warfare. They point to the use of the computer virus known as Stuxnet, which is believed to have caused damage to Iranian nuclear facilities in 2009, as an example of the potential and power autonomous cyber

[20] Myers (2015).

[21] United States Defense Advanced Research Projects Agency (2013). Note, however, that at least initially the vessel is designed to require human approval before launching an attack. The United States Navy is developing similar underwater systems to conduct de-mining operations; Ackerman (2013).

[22] Guarino (2013).

[23] Guarino (2013).

weapon systems hold.[24] Regardless of whether Stuxnet truly represents the world's first fully autonomous weapon system, it seems likely that cyberspace will continue to be a keen area of interest for autonomous systems research and development.

The exact path these advances in autonomous technology might take in the future is, however, uncertain. Many experts expect miniaturization to play a major role in future weapons development.[25] Based upon improvements in nanotechnology and other fields, future autonomous systems are envisioned to be more compact and be able to operate over substantially increased distances and time. These systems will not only be smaller but will be more expendable. Many of them will be designed to operate collectively as part of a swarm, capable of quickly and overwhelmingly attacking an enemy force.[26] The computing power supporting future autonomous systems will unquestionably be more powerful and faster than existing capabilities. Future systems may increasingly be controlled with sophisticated forms of artificial intelligence, known as general or strong artificial intelligence. These systems will have an enhanced capacity to independently respond and adjust to their environment.[27] While the promise of future technological advances is bright, one should not expect major breakthroughs to be the norm. Instead, it is more likely that innovations in autonomy will occur incrementally and gradually over time.[28] As the technology slowly improves, the systems will begin assuming greater control over their own actions. Militaries are likely to accept and seek such moves towards autonomy because of the enormous operational advantages inherent with such weapons.

3.2 *Operational Benefits of Autonomy*

Operational benefits may propel nations to increasingly seek autonomous systems. First, current remotely controlled systems may prove to be prohibitive in terms of personnel. These systems require significant numbers of personnel to pilot and maintain the systems as well as to constantly analyse the data they obtain. Militaries unable to meet the demand for such man-piloted systems will instead turn to autonomous ones to fill the gap. As an example, the United States, in a recent

[24]Healey (2013).

[25]Guarino (2013).

[26]United States Air Force (2009), p. 16 (stating that '[a]s autonomy and automation merge, [systems] will be able to swarm ... creating a focused, relentless, and scaled attack'). The United States Air Force's Proliferated Autonomous Weapons may represent an early prototype of future swarming systems. See Singer (2009), p. 232; Alston (2011), p. 43.

[27]Singer (2009), p. 74; Kellenberger (2011), p. 27. Note, consensus does not exist as to if and when general artificial intelligence might become available. Artificial intelligence has previously failed to live up to some expectations. Computer scientist Noel Sharkey doubts that artificial intelligence advances will achieve human-like abilities in even the next 15 years; Sharkey (2011), p. 140.

[28]Anderson and Waxman (2013), p. 2.

roadmap on unmanned systems, pledged in the future to 'take the "man" out of unmanned'.[29] Given that many of the duties will be performed automatically by computers, autonomous systems generally require far fewer human operators.[30] Additionally, the use of autonomous systems will likely also provide an overall increase in capacity because, unlike human operators, autonomous weapons do not need to sleep or eat and do not become distracted.

Second, autonomous weapon systems are generally less susceptible to electronic or cyber-attacks. As adversaries increasingly become competent in jamming communications and attacking computer networks, communications links represent critical vulnerabilities. Remotely controlled systems are often completely dependent upon a satellite tether back to a human pilot and are incapable of completing their mission if that communications link becomes severed. Autonomous systems, on the other hand, could conceivably continue operations unimpeded even in adverse and electronically contested areas. By not being reliant on a constant link with a human operator, autonomous weapons would also provide fewer windows of opportunity for cyber-attackers to hack into the systems. Future militaries may find such hardened systems particularly advantageous.

Third, only autonomous systems may be able to operate fast enough for the increased pace of combat in the future. Manned and tethered systems inherently suffer from communications and human operator reaction delays. These time lags may make such systems inferior to and ineffective against an adversary's autonomous weapons.[31] Many commentators contend that conflicts in the future will be waged in an environment 'too fast [and] ... too complex for humans to direct'.[32] To avoid facing such a potential competitive disadvantage, nations will invariably seek to enhance their development of autonomous weapons.[33] With such operational advantages to be gained, nations can be expected to make autonomous weapons an ever growing segment of their military arsenals.

---

[29]United States Department of Defense (2013), p. 25. Under a heading labelled 'A Look to the Future' it explains: 'Currently personnel costs are the greatest single cost in (the Department of Defense), and unmanned systems must strive to reduce the number of personnel required to operate and maintain the systems. Great strides in autonomy, teaming, multi-platform control, tipping, and cueing have reduced the number of personnel required, but much more work needs to occur.'

[30]'Enable humans to delegate those tasks that are more effectively done by computer ... thus freeing humans to focus on more complex decision making'; United States Department of Defense (2012b), p. 1.

[31]Sharkey (2012), p. 110.

[32]Singer (2009), p. 128.

[33]For example, the United States has expressed an interest in seeking an expansion of autonomous features, albeit not lethal targeting capabilities, into its systems in the future; United States Department of Defense (2012b), pp. 1–3; United States Department of Defense (2013), p. 25.

4 An Analysis of Autonomous Weapons and Feasible Precautions in Attack

Assuming that technology will make autonomous weapon systems possible and operational realities will cause nations to clamour for them, it is important to carefully consider what law would apply to these new means and methods of warfare. If these systems were to be employed in a situation involving an armed conflict, then the law of armed conflict would undeniably be the relevant body of law. A State hoping to develop and deploy an autonomous weapon must analyse two separate tracks of the law in detail.[34] First, a State must determine whether the weapon itself is lawful *per se*. Second, it would need to appraise whether the use of the weapon is prohibited. Only when a State confirms that the autonomous weapon fully complies with both facets of the law may the new weapon be employed. While questions exist for autonomous weapons with both tracks, the more significant and complex issues lie in this second track related to weapons use. Most acute among those use related concerns is the law of armed conflict duty to take all feasible precautions in attack.

As codified in Article 57 of Additional Protocol I, the rule, which reflects customary international law, requires an attacker to take multiple precautions to help minimize the harm to the civilian populace. The rule directs that an attacker exercise 'constant care' to shield civilians and civilian objects from the effects of attacks. The article further expands on what is required to meet these obligations by specifying that an attacker must (1) 'do everything feasible to verify that the objectives to be attacked are neither civilians nor civilian objects and are not subject to special protection but are military objectives'; (2) cancel an attack if it becomes apparent that the rules of distinction or proportionality will be breached; (3) provide 'effective advance warning' of an attack if it may affect the civilian population, 'unless circumstances do not permit'; (4) '[w]hen a choice is possible between several military objectives for obtaining a similar military advantage, [select] that the attack on which may be expected to cause the least danger to civilian lives and to civilian objects'; and (5) 'take all feasible precautions in the choice of means and methods of attack with a view to avoiding, and in any event to minimizing, incidental loss of civilian life, injury to civilians and damage to civilian objects'.[35] Each of these obligations applies to autonomous weapons equally as it does for any weapon system. The remainder of this section of the chapter explores how the unique aspects of autonomy will play a role with compliance of each of these legal obligations.

[34] *Legality of the Threat or Use of Nuclear Weapons (Advisory Opinion)*, ICJ Reports 1996, p. 226 (hereinafter *Nuclear Weapons*); Schmitt and Thurnher (2013), p. 243. Schmitt (2013b), p. 8.

[35] Additional Protocol I, Article 57.

4.1 Verify Objectives

The first feasible precaution to examine is the requirement to 'do everything feasible to verify that the objectives to be attacked are neither civilians nor civilian objects and are not subject to special protection but are military objectives'. The essence of this requirement is the need to follow the law of armed conflict principle of distinction, long considered a cardinal or bedrock principle.[36] Distinction obliges a combatant to distinguish between combatants and civilians, and also between military objectives and civilian objects. Codified in Article 48 of *Additional Protocol I* with accompanying rules in Articles 51 and 52,[37] the principle is also considered reflective of customary international law.[38] Undeniably, any autonomous weapon system must comply with this principle and be capable of being limited strictly to attacking only military objectives.

Critics who charge that autonomous weapons will not be able to comply with the principle of distinction,[39] fail to fully appreciate how the surrounding context and environment factor into this analysis. When examining the principle of distinction, it is critical to first determine the type of environment in which a State plans to use the autonomous weapon system. If, for example, an autonomous weapon was being deployed to a remote area generally devoid of civilians, such as underwater or a desert environment, then the demands on the weapon to distinguish would be much lower. In fact, in circumstances such as these, an autonomous weapon system could possibly comply with the rule even if it possessed only a low level ability to distinguish. As the battlefield environment gets more complicated and interspersed with civilians, the demands on the system raise considerably. In urban or other populated areas, an autonomous weapon would likely need to have advanced sensor packages and recognition software. Even then, it is conceivable that the battlefield situation might be too cluttered for the system to accurately distinguish between military objectives and civilian objects or between combatants and the civilian population. In those cases, an autonomous weapon would clearly be unlawful to use. Thus, the environment surrounding the use of the weapon plays a major role.

In order to meet their obligation to do everything feasible to verify the target when using autonomous weapons in an armed conflict, nations will likely need to use various sophisticated sensors to help ensure the reliability of the target identification. Many experts contend that sensor systems used for targeting autonomous weapons must be essentially equivalent to those used with manned or remotely piloted systems.[40] Additionally, it would be unlawful to use an autonomous weapons

[36]The International Court of Justice has recognised distinction as a 'cardinal' principle of the law of armed conflict. *Nuclear Weapons*, paras. 78–79.

[37]Additional Protocol I, Articles 49, 51–52.

[38]Henckaerts and Doswald-Beck (2005), r. 1; *Nuclear Weapons Case*, paras. 78–79; Cadwalader (2011), p. 157.

[39]See for example, Human Rights Watch (2012), pp. 30–32.

[40]HPCR (2009), r. 39.

by itself if other external systems, which could significantly help improve target identification, were militarily feasible to use under the given circumstances.

The key question in analysing whether additional systems need to be used to augment an autonomous weapon in the situation described above is what constitutes the term 'feasible'. Feasible has been equated to mean that which is 'practicable or practically possible, taking into account all circumstances ruling at the time, including humanitarian and military considerations.'[41] Military considerations take into account operational concerns such as the survivability of military personnel involved and the weapon system itself. An examination of feasibility is a question of reasonableness. An attacker is not required to assume greater risk to protect the civilian populace if a reasonable attacker in a similar situation would not take such actions. Therefore, when a nation examines whether it must augment the reliability of an autonomous weapon by using other external systems, that nation must determine if the degree of operational risk to the external systems is reasonable. If, for example, the external system could be used elsewhere against higher priority targets or in a different battlefield situation where it can better minimize harm to civilians, then its use with the autonomous weapon would not be considered feasible. This question of what is feasible in a given situation permeates all of the following discussions about precautions in attack rules, including when a nation must cancel an attack.

4.2 Suspend or Cancel Attack

The second precautions in attack rule requires an attacker to cancel or suspend an attack if it becomes apparent that the attack might violate either the principle of distinction or proportionality. The discussion about distinction from above applies to this provision as well. However, a potentially bigger legal concern may stem from the portion of the rule that deals with proportionality. The principle of proportionality is codified in both Article 51(5)(b) and Article 57(2)(a)(iii) of Additional Protocol I.[42] Reflective of customary international law,[43] this complex principle prohibits an attack if it is 'expected to cause incidental loss of civilian life, injury to civilians, damage to civilian objects, or a combination thereof, which would be excessive in relation to the concrete and direct military advantage anticipated.'[44] The first key component in this analysis is the expected amount of collateral damage. For the use of an autonomous weapon to be lawful, the weapon system would have to be able to produce an estimate of the number of civilians who may be harmed incidentally as a result of an attack. This step of the analysis

[41] Henckaerts and Doswald-Beck (2005), r. 15; HPCR (2009), r. 1(q).

[42] Additional Protocol I, Articles 51(5)(b), 57(2)(a)(iii).

[43] Henckaerts and Doswald-Beck (2005), r. 14; Cadwalader (2011), pp. 157–158.

[44] Additional Protocol I, Article 51(5)(b).

is unlikely to be a challenge for autonomous systems because it is essentially a quantitative determination. The United States and other major militaries have developed a system to make such estimates called the Collateral Damage Estimation Methodology, which relies on scientific data and objective standards.[45] The second step of the proportionality analysis involves more qualitative judgements and will present greater challenges for autonomous systems.

If the collateral damage estimate reveals that civilian casualties or damage to civilian objects are expected from an attack, then the attacking force is obliged to consider the amount of harm to civilians or damage to civilian objects in relation to the anticipated military advantage of the attack. Determining the relative value of the military advantage of an attack is both subjective and contextual. It is unlikely an autonomous system will be capable of making independent value judgements as required by the proportionality principle. That is not to say, however, that the use of autonomous weapons would automatically be unlawful in such situations, as several critics contend.[46] For instance, in circumstances where civilians are not expected to be harmed, this principle would not be in effect. At other times when civilians might be expected to be harmed in a potential attack, nations might be able to comply by instead relying on preprograming the acceptable values for the various situations that an autonomous weapon might encounter. The military operators would thus establish a framework for the autonomous system to operate within, whereby the operator has pre-determined for any specific target what the commander considers as an excessive amount of collateral damage. Although creating such a framework may be technologically challenging, it is not inconceivable that such a mechanism could be developed and embedded into an autonomous weapon. To ensure compliance with the principle, one can expect any predetermined values to initially be set conservatively. Moreover, militaries will likely resort to controlling autonomous weapons tightly with geographic boundaries or time limits as an additional compliance measure.

One additional point to consider is that this obligation to suspend or cancel an attack represents a continuing duty to comply. Nations intending to deploy autonomous weapons must ensure that they have protections in place to ensure a proposed attack with an autonomous system can and will be stopped if they learn that the circumstances on the ground have drastically changed. While solving these dilemmas will pose difficulties for nations intending to use autonomous weapons, it is important to remember that these types of proportionality and precautions decisions are complex even for human commanders. Autonomous weapons will face similar challenges and must at least meet the standard set by humans. If States can find a reasonable solution to these proportionality and precautions requirements, then their use of autonomous weapons should not be deemed unlawful.

---

[45]For a discussion of the collateral damage methodology used by the United States military, see Thurnher and Kelly (2012).

[46]For example, Human Rights Watch maintains that an autonomous weapon 'could not be programmed to duplicate the psychological processes in human judgment that are necessary to assess proportionality.' Human Rights Watch (2012), p. 33.

4.3 Provide Effective Warning

The third precautions in attack rule is one that requires an attacker provide 'effective advance warning' of an attack if it may affect the civilian population, 'unless circumstances do not permit'.[47] There are additional obligations to warn certain medical vessels or facilities that forfeit their protected status. Nations will need to consider how warnings can be given whenever autonomous weapons are deployed into an armed conflict.

There are no established standard forms for the warnings.[48] The warnings can be a general message delivered either to the leadership of the enemy nation or directly to the civilian population. An attacker is not required to explicitly detail the particular time or place of the planned attack, but the warning should provide as much detail as the circumstances allow.[49] The warning must be sufficient to provide the civilian population the opportunity to take measures to avoid the dangers. The delivery method of the warning can vary and can range from methods such as leaflets or media broadcasts.[50]

As spelled out in the plain language of the rule, the obligation to issue a warning only exists if and when the civilian populace will be affected by the attack. If there is no expected harm to civilians or civilian objects, then an attacker need not provide a warning. Furthermore, the rule provides an exception that would allow an attacker to commence an attack without issuing a warning, even if the attack may be expected to harm civilians. The exception exists when the circumstances do not permit the issuing of a warning. The acceptable circumstances can include legitimate military factors. For instance, a warning would not be required if the warning would undermine the operational surprise needed to accomplish the mission or if the attacking forces would face greater risk after the issuance of a warning.

Nations desiring to deploy autonomous weapons must still contend with the requirement to issue effective warnings to the civilian populace as long as circumstances allow. These nations will need to take several measures to ensure compliance. The systems will need to be designed to identify when the civilian population might be expected to be harmed by an attack. As was mentioned above in the Collateral Damage Estimation Methodology discussion, it seems likely that autonomous weapons will have adequate sensor and computational capabilities to determine whether the civilian populace can be expected to be harmed in an attack. If collateral damage is to be expected, then the system needs to know whether the circumstances dictate issuing a warning or relying on the exception. Nations will need to determine how best to answer this highly contextual and arguably subjective question. If the circumstances permit, and therefore require, the issuing of a

[47] Additional Protocol I, Article 57(2)(b).

[48] United States Department of Defense (2015), r. 5.11.

[49] HPCR (2009), r. 38.

[50] United States Department of Defense (2015), r. 5.11.

warning, then nations will need to determine how best to make an effective warning. It seems reasonable to assume that nations could design autonomous systems to have broadcast or loudspeaker-type capabilities. Alternatively, the autonomous system could be set to refrain from attacking until it received confirmation that a warning was effectively made by some other type of system.

One additional consideration in this analysis with regard to autonomous systems and warnings is that autonomous systems do not necessarily have the same need for self-preservation as manned systems. As a result, it may potentially be easier for a nation attacking with autonomous weapons to comply with this rule. For instance, if the attacking nation had a large number of autonomous vehicles participating in an attack, as may be expected with so-called swarm technology, then the attacking nation may have greater abilities to risk some of the systems in an effort to affect the warning. Regardless, nations must understand this obligation and consider how best to issue required warnings when using autonomous weapons. Similarly, nations must understand the obligation to choose the objective which may be expected to cause the least danger to the civilian populace.

4.4 Choose Between Objectives

The fourth precaution in attack rule requires an attacker to select the object of attack that is expected to cause the least danger to civilians whenever 'a choice is possible between several military objectives for obtaining a similar military advantage'. At least one leading nation has expressed doubt as to whether this provision is truly reflective of customary international law.[51] If the provision is however assumed to reflect customary international law, then this requirement would apply equally to those launched by autonomous weapons. Selecting between different military objectives may present challenges for autonomous weapons.

One classic example of how this obligation might unfold is as follows. An attacker cannot lawfully use force against an electrical substation if it was militarily feasible to instead attack the electronic transmission lines and achieve the same military advantage while endangering fewer civilians. When analysing whether attacking one of the targets is feasible, an attacker can and should consider military factors such as how significant the risk is to their own or friendly forces and whether they have adequate assets available to conduct the attack against the particular target. For instance, if one of the targets is more heavily defended, the attacker need not choose that target even if less collateral damage may result. Attacking the more heavily defended location would instead be considered militarily unreasonable. Additionally, the term 'similar military advantage' is understood to

[51]For example, the United States has issued statements challenging the notion that this provision of Additional Protocol I reflects customary international law. United States Department of Defense (2015), r. 5.11.5.

be the anticipated military advantage of the attack taken as a whole, rather than subdivided into parts. Given that the analysis focuses on selecting one target over the other, it is essential that the military advantage anticipated from each possible target be considered as a whole rather than in isolation.[52]

These decisions involve complex, contextual judgments. It may be challenging to embed such qualitative decision making capabilities into an autonomous weapon system. While commanders may be able to predict different scenarios and situations the autonomous weapon may face and proactively provide input to the weapon system's direction control mechanisms, this area of the law must be carefully considered and resolved by nations seeking to employ autonomous weapons. Similar challenges also exist when one looks to the final precaution in attack rule.

4.5 *Choose Means and Methods with a View to Minimizing Harm to Civilians*

The fifth precaution in attack rule is the obligation of attackers to 'take all feasible precautions in the choice of means and methods of attack with a view to avoiding, and in any event to minimizing, incidental loss of civilian life, injury to civilians and damage to civilian objects'.[53] This final precaution requires nations to select the means and methods of warfare expected to cause the least damage to the civilian populace without sacrificing military advantage. Means equates to weapons and weapon systems, whereas methods refers to military tactics, techniques, and procedures. Nations seeking to develop autonomous weapons will need to consider the best methods for complying with this rule. If an autonomous weapon possesses multiple munitions onboard, then it must be able to calculate which warheads or weapons can accomplish the mission. It will further need to be able to determine the amount of harm to the civilian populace each of those weapons would be expected to produce. Then it must select the one that is expected to cause the least harm. Given the projected advances in technology and the likely ability of systems in the future to calculate collateral damage estimates on their own, it can be expected that systems will be able to adequately make these decisions without further human operator involvement.

Autonomous weapons may also be able to make decisions about when and how to attack certain objectives in a way to help reduce the risk of incidental harm. One method might be to adjust the timing of an attack to do so when civilians are less likely to be present in a particular area.[54] An autonomous weapon might be programmed to refrain from attacking a moving enemy until the enemy has moved to an isolated or less populated area. An autonomous weapon might also

[52]HPCR (2009), r. 33.

[53]Additional Protocol I, Article 57.

[54]United States Department of Defense (2015), r. 5.11.2.

be programmed to understand complex military weaponeering, or the process of determining the quantity and other characteristics of a weapon required to achieve a certain effect on the battlefield. Embedded with this knowledge, autonomous weapons may be able to determine what angles of attack might cause the least likely harm to surrounding civilians, or they may be able to understand when to arm a weapon with time fuze delays or other mitigation techniques to reduce civilian casualties. This ability to take steps toward selecting means and methods with a view to minimizing harm to the civilian populace seems conceivable in the future given expected technological advances. It will need to be embedded into any future autonomous weapon, which might encounter civilians on the battlefield.

Looking at this obligation more broadly, one can identify how the rule contains significant implications for the controversy surrounding the lawfulness of autonomous weapons. If an autonomous weapon can be expected to cause more harm to the civilian populace than a human controlled system, and the human controlled system can be used without sacrificing military advantage or unreasonably increasing the risk to the human operator, then, as a matter of law, the autonomous weapon could not be lawfully used on the battlefield. In essence, an autonomous weapon system can only be used when it might achieve the same military advantage yet cause less damage to civilians and civilian objects than other available and feasible systems.

Naturally there is a fair degree of flexibility in applying this rule. After all, the analysis is based upon determining what is feasible or practicable. That said it is an important point worth noting in the ongoing debate about whether autonomous weapons should be pre-emptively banned. By complying with this tenet of the law of armed conflict, states will essentially be ensuring that autonomous weapons will only be used when other systems similarly capable of accomplishing the mission, and doing so in a manner expected to cause less collateral damage, are unavailable. Opponents of autonomous weapons often overlook the implication of this rule and fail to realize the potential negative impact of a ban. If a ban were indeed enacted, then militaries in the future might possibly miss an opportunity to field a system, which could conceivably possess superior decision making abilities or more precise sensor packages. As a result, they might be forced to rely on a system that causes more danger to civilians, assuming the system comports with the rule of proportionality. Civilians and civilian objects would thus be placed in greater risk than had the attacker been able to utilize the autonomous weapon system.

5 Conclusion

Whereas human judgement is often clouded by anger, revenge, fear, and fatigue, autonomous machines may offer the promise of removing those worries from the battlefields of the future. Nations are encouraged by this prospect and are actively seeking to embed autonomous properties into their military systems. This trend toward greater levels of autonomy in weapon systems can be expected to continue.

Technology may soon make possible the creation of autonomous weapon systems operating within tightly controlled parameters but able to select and engage targets without human intervention. Such systems will not only help protect a nation's own fighting forces from danger but may also potentially provide better safeguards for civilian populations through more precise and accurate targeting.

These emerging systems have nevertheless become the subject of tremendous criticism. As with the creation of any new means of warfare, some controversy is to be expected. Calls for bans on new weapon systems are common throughout history. From the crossbow to aerial bombers, critics have repeatedly argued that various new weapons undermine the then existing laws of war. In this case, the proposed ban is unnecessary and unwise. The law of armed conflict properly represents a balance between the cardinal principles of military necessity and humanity. It contains sufficient protections for civilians while allowing for the accomplishment of the military mission. The law will adapt and evolve as needed to the challenges autonomous systems present, but, in principle, the law of armed conflict and autonomy are not incompatible. This is one of the reasons the International Committee for the Red Cross has thus far not supported a ban on autonomous weapon systems. Instead, they wisely call on nations developing autonomous systems to remain mindful of the tenets of the law of armed conflict and only develop systems that fully comply with its provisions.[55] Proponents of a ban on autonomous systems seemingly fail to appreciate the significance of the law of armed conflict's existing protections.

When examining autonomous weapon systems from a law of armed conflict perspective, as this chapter has sought to do, one will determine that autonomous weapons will likely be deemed lawful in many scenarios. The use of autonomous weapon systems may be limited in some complex battlefield situations because of the obligation to take all feasible precautions in attack. In other cases, however, autonomous weapons should be able to fully comply with those principles given the particulars of the battlefield environment. Moreover, as technology improves, one would expect the systems to be capable of being used in an increasing number of circumstances.

States are closely following the controversy surrounding autonomous weapon systems, and are sensitive to the concerns. They recognise the need to craft policies and targeting procedures that ensure full compliance with the law of armed conflict, and a number of States have already implemented some initial measures.[56] As the development of autonomous systems continues, States will need to adjust their targeting procedures to properly consider the unique issues autonomy raises with respect to taking precautions in attack. Given the extraordinary potential benefits of these systems, States have huge incentives to work through these difficult challenges. Leading military States should highlight, through actions and words,

[55] International Committee of the Red Cross (2013).

[56] The United States issued a policy directive in 2012 establishing a strict approval process for any AWS acquisitions or development and mandating various safety measures be incorporated into future AWS designs. United States Department of Defense (2012c).

how their respect for the law of armed conflict is constraining their development of these news weapons. If that begins to occur, it seems likely that States will coalesce around the ideas that the law of armed conflict meets the challenges of this emerging technology and a ban on autonomous weapons will be unnecessary.

References

Ackerman S (2013) Navy preps to build a robot ship that blows up mines. www.wired.com/dangerroom/2013/01/robot-mine-sweeper/. Accessed 29 Sept 2015

Alston P (2011) Lethal robotic technologies: the implications for human rights and international humanitarian law. J Law Inf Sci 21:35–60

Anderson K, Waxman M (2013) Law and ethics for robot soldiers: why a ban won't work and how the laws of war can. Hoover Inst Policy Rev

Cadwalader G (2011) The rules governing the conduct of hostilities in additional protocol I to the Geneva conventions of 1949: a review of relevant United States references. Yearb Int Humanitarian Law 14:133–171

Campaign to Stop Killer Robots (2013) Who we are. http://www.stopkillerrobots.org/coalition. Accessed 29 Sept 2015

Convention on Conventional Weapons (2013) Final report of the meeting of the high contracting parties to the Convention on Prohibitions or Restrictions on the Use of Certain Conventional Weapons Which May Be Deemed to Be Excessively Injurious or to Have Indiscriminate Effects, CCW/MSP/2013/CRP.1

Dahm W (2012) Killer drones are science fiction. Wall Street J. 15 Feb 2012

Guarino A (2013) Autonomous cyber weapons no longer science-fiction, Engineering and Technology Magazine online. http://eandt.theiet.org/magazine/2013/08/intelligent-weapons-are-coming.cfm. Accessed 29 Sept 2015

Harvard Program on Humanitarian Policy and Conflict Research (HPCR) (2009) Manual on international law applicable to air and missile warfare

Healey J (2013) Stuxnet and the dawn of algorithmic warfare, Huffington Post Online. http://www.huffingtonpost.com/jason-healey/stuxnet-cyberwarfare_b_3091274.html. Accessed 29 Sept 2015

Henckaerts J, Doswald-Beck L (eds) (2005) International committee of the red cross customary international humanitarian law. Cambridge University Press, Cambridge

House of Lords Debate 26 March 2013 (Lord Astor of Hever, Parliamentary Under Secretary of State, Defence). http://www.publications.parliament.uk/pa/ld201213/ldhansrd/text/130326-0001.htm#st_14. Accessed 29 Sept 2015

Human Rights Watch (2012) Losing humanity: the case against killer robots. www.hrw.org/sites/default/files/reports/arms1112ForUpload_0_0.pdf. Accessed 29 Sept 2015

International Committee for the Red Cross (2013) Autonomous weapons: states must address major humanitarian, ethical challenges. http://www.icrc.org/eng/resources/documents/faq/q-and-a-autonomous-weapons.htm. 29 Sept 2015

Kellenberger J (2011) Keynote address. In: von Heinegg WH, Beruto GL (eds) International humanitarian law and new weapon technologies. International Institute of Humanitarian Law, Sanremo, pp 23–27

Krishnan A (2009) Killer robots: legality and ethicality of autonomous weapons. Ashgate, Burlington

Myers M (2015) Unmanned X-47B Jet Completes First Aerial Refueling, Navy Times Online. http://www.navytimes.com/story/military/2015/04/22/navy-navair-x-47b-unmanned-jet-refueling/26191213/. Accessed 3 Oct 2015

Poitras C (2012) Smart robotic drones advance science. http://today.uconn.edu/blog/2012/10/smart-robotic-drones-advance-science/. Accessed 29 Sept 2015

Public Broadcasting Service (2011) Smartest machines on earth. (transcript) www.pbs.org/wgbh/nova/tech/smartest-machine-on-earth.html. Accessed 29 Sept 2015

Russell S, Norvig P (2010) Artificial intelligence: a modern approach, 3rd edn. Prentice Hall, Upper Saddle River

Schmitt MN (2013a) Autonomous weapon systems and international humanitarian law: a reply to the critics. Harv Natl Secur J Features. http://harvardnsj.org/wp-content/uploads/2013/02/Schmitt-Autonomous-Weapon-Systems-and-IHL-Final.pdf. Accessed 29 Sept 2015

Schmitt MN (ed) (2013b) Tallinn manual on the international law applicable to cyber warfare. International group of experts at the invitation of the NATO Cooperative Cyber Defence Centre of Excellence. Cambridge University Press, Cambridge

Schmitt MN, Thurnher J (2013) Out of the loop': autonomous weapon systems and the law of armed conflict. Harv Natl Secur J 4:231–281

Sharkey N (2011) Automating warfare: lessons learned from the drones. J Law Inf Sci 21:140–154

Sharkey N (2012) Drones proliferation and protection of civilians. In: Heitschel von Heinegg W (ed) International humanitarian law and new weapon technologies. International Institute of Humanitarian Law, Sanremo, pp 108–118

Singer P (2009) Wired for war: the robotics revolution and conflict in the twenty-first century. Penguin Press, New York

Thurnher J, Kelly T (2012) Collateral damage estimation, US Naval War College video. www.youtube.com/watch?v=AvdXJV-N56A&list=PLam-yp5uUR1YEwLbqC0IPrP4EhWOeTf8v&index=1&feature=plpp_video. Accessed 29 Sept 2015

United Nations (2013) Report of the special rapporteur on extrajudicial, summary or arbitrary executions, Christof Heyns. UN Doc A/HRC/23/47

United States Air Force (2009) Unmanned aircraft systems flight plan 2009–2047. Headquarters Department of the Air Force, Washington

United States Defense Advanced Research Projects Agency (2013) DARPA's anti-submarine warfare game goes live. www.darpa.mil/NewsEvents/Releases/2011/2011/04/04_DARPA's_Anti-Submarine_Warfare_game_goes_live.aspx. Accessed 29 Sept 2015

United States Department of Defense (2012a) Directive 3000.09: Autonomy in weapon systems. Government Printing Office, Washington

United States Department of Defense (2012b) Directive 3000.09: Autonomy in weapon systems: response-to-query talking points. Government Printing Office, Washington (on file with author)

United States Department of Defense (2012c) Task force report: the role of autonomy in DoD systems. www.fas.org/irp/agency/dod/dsb/autonomy.pdf. Accessed 29 Sept 2015

United States Department of Defense (2013) FY2013–2038 Unmanned systems integrated roadmap. Government Printing Office, Washington

United States Department of Defense (2015) Law of War Manual, Office of General Counsel. http://www.dod.mil/dodgc/images/law_war_manual15.pdf

Von Heinegg WH (2011) Concluding remarks. In: von Heinegg WH, Beruto GL (eds) International humanitarian law and new weapon technologies. International Institute of Humantiarian Law, Sanremo, pp 183–186

Unmanned Maritime Systems: Does the Increasing Use of Naval Weapon Systems Present a Challenge for IHL?

Wolff Heintschel von Heinegg

Abstract The legal status of unmanned maritime systems is unsettled. Whereas this does not pose insurmountable problems in times of peace, it could prove as an obstacle to the use of unmanned maritime systems for the exercise of belligerent rights in times of international armed conflict. Nevertheless, unmanned maritime systems will qualify as means of warfare, if they are used for attack purposes. While they are not unlawful per se, even if semi-autonomous or autonomous, their use will have to be in compliance with targeting law.

1 Systems and Vehicles

Naval mines and torpedoes could well qualify as unmanned maritime systems. Still, they are excluded here because

- they are regulated by the 1907 Hague Convention VIII and
- because they normally are not recoverable or designed to return (you do not want a torpedo or a naval mine to return to its origin).

The terms 'unmanned maritime systems' (UMS) and 'unmanned maritime/seagoing vehicles' (UMVs/USVs) seem to indicate that the former consist of various components and are far more complex than the latter. Indeed, 'system' means a 'complex whole' or a 'set of things working together as a mechanism or interconnecting network'.[1] According to the U.S. DoD,

> UMS comprise unmanned maritime vehicles (UMVs), which include both unmanned surface vehicles (USVs) and unmanned undersea vehicles (UUVs), all necessary support

W. Heintschel von Heinegg, Professor, Chair for Public Law, International Law, European Law and Foreign Constitutional Law, European University Viadrina, Frankfurt (Oder), Germany.

[1]Concise Oxford English Dictionary (2011), p. 1462.

W. Heintschel von Heinegg (✉)
Chair for Public Law, International Law, European Law and Foreign Constitutional Law, European University Viadrina, Frankfurt (Oder), Germany
e-mail: heintschelvonheinegg@europa-uni.de

W. Heintschel von Heinegg et al. (eds.), *Dehumanization of Warfare*,
DOI 10.1007/978-3-319-67266-3_7

components, and the fully integrated sensors and payloads necessary to accomplish the required missions.[2]

Although those definitions seem to suggest that UMVs/USVs are but components of UMS it would not be correct to hold that UMVs/USVs do not qualify as 'systems' because they are composed of various subsystems.[3] Since a distinction between 'systems' and 'vehicles' does not prove helpful it seems to be correct to consider the terms 'UMS' and 'UMV' as widely synonymous.

Accordingly, UMS are self-propelled or remotely-navigated craft that are normally recoverable and designed to perform certain functions at sea by operating on the surface, semi-submerged or undersea. UMS are remotely operated, remotely controlled/supervised or they perform some or all of their functions independently from a human controller or operator.

2 Missions/Tasks of UMS

UMS can perform a wide variety of missions or tasks. While the focus of this paper is on military uses it is important to bear in mind that UMS, in particular UUVs, are today used for the performance of the following important civilian/non-military tasks:

1. Offshore oil and gas missions;
2. undersea cable deployment and inspection;
3. commercial salvage;
4. aquaculture; and
5. science missions, such as oceanography and marine archaeology.[4]

According to the DoD Roadmap, current military missions performed by UMS (USVs and UUVs) include 'mine warfare, mine neutralization, reconnaissance, surveillance, hydrographic surveying, environmental analysis, special operations, and oceanographic research'.[5] Similarly, the *UUV Master Plan* identifies nine specific mission categories and prioritizes them as follows[6]: (1) Intelligence, surveillance, and reconnaissance (ISR); (2) mine countermeasures (MCM); (3) anti-submarine Warfare (ASW); (4) inspection/identification; (5) oceanography;

[2]U.S. Department of Defense (FY 2013–2038), p. 8.

[3]For example, the major UUV's subsystems are: the pressure hull, the hydro-dynamic hull, ballasting, power and energy, electrical-power distribution, propulsion, navigation and positioning, obstacle avoidance, masts, maneuver control, communications, locator and emergency equipment, payloads. See National Defense Research Institute (2009), p. 46 et seq.

[4]National Defense Research Institute (2009), p. 41 et seq.

[5]U.S. Department of Defense (FY 2013–2038), p. 109.

[6]U.S. Department of the Navy (2004), p. 16.

(6) communication/navigation network node (CN3); (7) payload delivery; (8) information operations (IO); and (9) time-critical strike (TCS).[7]

3 Legal Status and Navigational Rights in General

Although UMS navigate in sea areas they cannot without difficulties be considered ships. The international law of the sea lacks a uniform definition of the term ‘ship’. UNCLOS[8] uses the terms ‘vessel’ and ‘ship’ interchangeably, without providing a definition of either term. In other treaties a ‘ship’ is defined as either (1) ‘any sea-going vessel of any type whatsoever, including floating craft, whether self-propelled or towed by another vessel, making a sea voyage’[9]; (2) ‘a vessel of any type whatsoever operating in the marine environment [. . .] includ[ing] hydrofoil boats, air-cushion vehicles, submersibles, floating craft and fixed or floating platforms’[10]; (3) ‘every description of water craft, including non-displacement craft and seaplanes used or capable of being used as a means of transportation on water’[11]; or (4) ‘any self-propelled sea-going vessel used in international seaborne trade for the transport of goods, passengers, or both’.[12] While those treaties do not prohibit treating UMS as vessels or ships in particular UNCLOS is designed for manned systems. In view of these difficulties UMS are not characterized as ships or vessels but rather as ‘craft’.[13] It is quite probable that a considerable number of States is not prepared to either recognize as or to assimilate UMS to ships/vessels. Still, for the purposes of the present manual UMS could be considered as vessels. This should be accompanied by a recommendation to governments to agree on a joint statement to that effect. That would render a formal modification or amendment of the existing international law unnecessary. Moreover, it would be possible to characterize UMS as warships, if they are operated by the armed forces of a State for exclusively non-commercial governmental purposes.

[7]For a detailed description of those missions/tasks see National Defense Research Institute (2009), p. 13 et seq.

[8]United Nations Convention on the Law of the Sea of 10 December 1982, 1833 UNTS 3, 397.

[9]1962 Amendments to the 1952 Convention for Prevention of Pollution of the Sea by Oil, 11 April 1962, Annex, Article 1(1), 600 UNTS 332.

[10]International Convention for Prevention of Pollution from Ships of 1973, as amended by the 1978 Protocol of 17 February 1978, 1340 UNTS 61, 184.

[11]Convention on the International Regulations for Preventing Collisions at Sea.

[12]United Nations Convention on Conditions for Registration of Ships (not in force), Article 2(4).

[13]U.S. Navy/U.S. Marine Corps/U.S. Coast Guard, The Commander’s Handbook on the Law of Naval Operations (NWP 1-14M), paras. 2.3.4–2.3.6 (Edition July 2007).

3.1 Sovereign Immunity

If UMS are operated by the armed forces or any other government agency of a State, they may not necessarily qualify as warships or State ships, but in view of the fact that they either constitute State property or serve exclusively non-commercial governmental functions, they do enjoy sovereign immunity and may be interfered with by other States in very exceptional circumstances (e.g. in an international armed conflict) only. Accordingly, it is correct to hold: 'USVs and UUVs engaged exclusively in government, non-commercial service are sovereign immune craft.'[14]

It is important to note that an independent legal status of sovereign immunity applies to UMS operating independently from another platform. Therefore, it is correct to hold that 'USV/UUV status is not dependent on the status of its launch platform.'[15] If the UMS is tethered to a controlling platform, it is difficult to attach to it an independent legal status. However, in view of the distinction between ROVs and AUVs becoming increasingly obsolete, the legal status of UMS should not necessarily be based on the controlling platform.

3.2 Navigational Rights

According to NWP 1-14M, 'USVs and UUVs retain independent navigation rights'[16] and they, thus, are considered as enjoying the same navigation rights as surface vessels and submarines. Other States have not (yet) given statements to that effect although they in fact make use of UMS for governmental, scientific and commercial purposes. Hence, it is safe to conclude that UMS enjoy the right of freedom of navigation in the high seas and in the EEZ as well as the rights of innocent passage, transit passage and archipelagic sea lanes passage.

4 UMS and IHL

4.1 Exercise of Belligerent Rights

Although it may be unsettled whether UMS qualify as, or are assimilated to, warships State practice seems to suggest that they are, and will be, used not only for attack purposes but also for the exercise of other belligerent rights, such as inspection of vessels. Since the exercise of those belligerent rights will

[14]NWP 1-14M (2007), para. 2.3.6.

[15]Ibid.

[16]Ibid., para. 2.5.

predominantly occur in high seas areas, there is a need for transparency because not only enemy vessels may be affected but also neutral vessels. Therefore, in times of international armed conflict UMS should be identifiable as belonging to the armed forces of a belligerent. Moreover, those controlling or pre-programming them should be under regular armed forces discipline in order to ensure compliance with the law of armed conflict.

4.2 *UMS as Means of Warfare*

Whereas many UMS are used for ISR or oceanography, some are designed for attack purposes, such as those employed for ASW, MCM or mine-laying. If and to the extent UMS are employed for the purposes of attack, they qualify as means of warfare[17] and their employment is subject to weapons law and targeting law.

4.2.1 Weapons Law

According to Article 35(2) AP I, it is prohibited to employ means or methods of warfare of a nature to cause superfluous injury or unnecessary suffering. According to Article 35(3) AP I, it is prohibited to employ methods and means of warfare which are intended or may be expected to cause widespread, long-term and severe damage to the natural environment. According to Article 51(4) lit. (b) and (c) AP I, it is prohibited to employ a means of combat which cannot be directed at a specific military objective or whose effects cannot be limited as required by AP I. Those prohibitions are customary in nature[18] but not equally relevant for the employment of current UMS as means of warfare.

As regards the capability of being directed at a specific military objective, it must be borne in mind that UMS are equipped with weapons that use the same or similar technologies as, for instance, modern naval mines or torpedoes. They carry weapons that home into targets that have been identified by magnetic, electromagnetic or other signatures allowing for a sufficiently reliable identification as lawful military objectives.

Regularly, the effects of current naval weapons systems will not be excessive in relation to the military advantage anticipated. In this context it is important to take into consideration the environment in which they are employed, in particular the characteristics of saline waters, and the fact that they will in most cases be used against targets which are not surrounded by civilian or specially protected objects. Apart from that, such considerations are not relevant under weapons law but rather under the law of targeting. The same holds true with regard to the protection of

[17] As defined in Rule 1 (t) AMW Manual.

[18] See Rules 5, 88, 89 AMW Manual.

the marine environment. No naval weapons system in existence is to be expected to inflict to the marine environment the kind of damage prohibited by Article 35(3) AP I.

It is therefore safe to conclude that UMS are lawful means of warfare, whether they are remotely operated/controlled or whether they operate with a certain degree of autonomy. However, UMS that operate with some autonomy must be capable of employment in accordance with the targeting law rules referred to in the next sub-section, and these will include the precautions in attack obligations set forth there.

4.2.2 Targeting Law

If UMS are employed for attack purposes, the same rules apply as in the case of the use of manned platforms qualifying as means of warfare. The target must be a lawful military objective, collateral damage may not be expected to be excessive in relation to the military advantage anticipated and, finally, the required precautions in attack must be taken.

With regard to the latter it may be asked whether at sea different standards apply. According to Article 57(4) AP I, military operations at sea are subject to 'all reasonable precautions'. Irrespective of the question whether the difference between the standards of feasibility and reasonableness is to be considered a 'tenuous nuance',[19] it follows from the wording of Article 57(4) AP I that the standard of reasonableness only applies to naval operations that may have an effect on the civilian population or civilian objects on land without being directed against targets on land. Sea-to-sea, as sea-to-air and air-to sea, operations, whose effects may not extend to land, continue to be governed by the customary obligation to take precautions in attack, which include the feasibility standard.[20] Accordingly, and without prejudice to the duty of constant care,[21] the employment of UMS for purposes of attack must be in accordance with the following requirements:

- The system and/or operator target must be able to verify the legality of the target[22]
- The system and/or operator must be capable of determining the probability of collateral damage[23]
  - If collateral damage is expected to be excessive in relation to the military advantage anticipated, the attack may not be executed[24]

[19]Sandoz et al. (1987), pp. 687 f., para. 2230.

[20]San Remo Manual, para. 46; AMW Manual, Rules 30 to 39.

[21]Article 57(1) AP I.

[22]Article 57(2) lit. (a)(i) AP I; San Remo Manual, para. 46(b).

[23]San Remo Manual, para. 46(a).

[24]Article 57(2) lit. (a)(iii) AP I; San Remo Manual, para. 46(d).

– If not excessive but still to be expected, there is an obligation to minimize or avoid collateral damage by all feasible precautions[25]
– Unless circumstances do not permit, an advanced effective warning must be given[26]

The obligation to avoid or minimize collateral damage by the choice of a different weapon ('weaponeering') will regularly have no effect because of the feasibility standard, because UMS do not, and probably will not, carry more than one type of weapon. The obligation to give an advanced warning will come into effect, if UMS are employed for mine-laying purposes, in particular within the EEZ of a neutral State.

4.2.3 Law of Neutrality

In times of an international armed conflict, the employment of UMS is subject to the law of maritime neutrality (unless the use of force has been authorized by a Chapter VII decision of the UN Security Council[27]). It is important to emphasize that it the law of neutrality applies to all military UMS whether they are used for attack purposes or not.

Accordingly, UMS may not engage in hostile actions in neutral waters or use neutral waters as a sanctuary or base of operations.[28] In the EEZ of a neutral State, they must be employed with due regard for the rights and duties of the coastal State.[29] The due regard rule also applies to the hostile use of UMS on the high seas.[30]

Belligerent UMS have the rights of transit passage and archipelagic sea lanes passage.[31] Accordingly, neutral States may not suspend, hamper, or otherwise impede those rights.[32] Belligerent UMS enjoy the right of innocent passage, unless the neutral coastal State has, on a non-discriminatory basis, conditioned, restricted or prohibited passage through its territorial sea.[33] The right of non-suspendable innocent passage may not be suspended.[34]

The parties to an international armed conflict must respect the sovereign immunity of neutral UMS, which are employed for non-commercial governmental

[25]Article 57(2) lit. (a)(ii) AP I; San Remo Manual, para. 46(c).

[26]Article 57(2) lit. (c) AP I.

[27]San Remo Manual, para. 7; AMW Manual, Rule 165.

[28]San Remo Manual, paras. 14–16.

[29]San Remo Manual, para. 34.

[30]San Remo Manual, para. 36.

[31]San Remo Manual, para. 28.

[32]San Remo Manual, para. 29.

[33]San Remo Manual, para. 19.

[34]San Remo Manual, para. 33.

purposes. Neutral civilian UMS may only be interfered with, if the qualify as lawful military objectives. It needs to be stressed that the capture of neutral civilian UMS under prize law will, if at all, apply in very exceptional situations only. Usually, UMS are not used for the transport of cargo that could constitute contraband.

5 Concluding Remarks

The fact that UMS are already in use for a variety of purposes and that it is highly probable that they will be increasingly used in the future, does not seem to pose a challenge to IHL. If UMS qualify as means of warfare, i.e. if designed or used for attack purposes, they are subject to weapons law and targeting law. They are not unlawful per se, and employment for attack purposes will regularly by in accordance with targeting law.

The only unresolved issue pertains to their status as warships, as defined by international law, and their entitlement to the exercise of belligerent and navigational rights.

References

Concise Oxford English Dictionary (2011) 12th edn. Oxford University Press, Oxford

Doswald-Beck L (ed) (1995) San Remo manual on international law applicable to armed conflicts at sea and explanation. Cambridge University Press

National Defense Research Institute (RAND Corp.) (2009) A Survey of Missions for Unmanned Undersea Vehicles. Available at: http://www.rand.org/content/dam/rand/pubs/monographs/2009/RAND_MG808.pdf

Program on Humanitarian Policy and Conflict Research, Manual on International Law Applicable to Air and Missile Warfare (2009) [AMW Manual]

Sandoz Y, Swinarski C, Zimmermann B (eds) (1987) Commentary on the additional protocols of 8 June 1977 to the Geneva conventions of 12 August 1949. Martinus Nijhoff, Leiden

U.S. Department of Defense, Unmanned Systems Integrated Roadmap FY 2013–2038

U.S. Department of the Navy (2004) The Navy Unmanned Undersea Vehicle (UUV) Master Plan

U.S. Navy/U.S. Marine Corps/U.S. Coast Guard, The Commander's Hand-book on the Law of Naval Operations (NWP 1-14M) (Edition July 2007)

Part IV
Cyber Warfare

Digital Forensics

Stefan Meier

Abstract In recent years, digital attacks against organizations, critical infrastructure and military targets are increasing. Generally, these attacks are summarized under the term Cyberwar and broadly discussed by the press, military experts and politicians. Attribution ("Who did it?") is often a major question in these discussions. But, by using computers and the Internet, these attacks leave digital traces which may become digital evidence. Digital forensics as a scientific discipline deals with methodologies to find and handle digital evidence. The main goal of digital forensic investigations is to reconstruct how an attack occurred and who is responsible. In this paper we show up the fundamental principles of digital forensics and discuss the usefulness of digital evidence in the mentioned attack scenarios. By sketching out two concrete examples, we present the capabilities of digital forensics to investigate distributed denial of service attacks and malware attacks.

1 Introduction

Through the ever-increasing digitization of all aspects of life, information systems' importance has rocketed over the past decades. Nowadays it is not imaginable how business or the private life would be without the support of information systems. However, information systems' security has fallen behind. Criminals, intelligence agencies and the military have realized the enormous potential of digital attacks. Targeted attacks against organizations have taken place all over in the last decade. States have been brought offline.[1] Physical machines have been damaged remotely.[2] Isolated facilities have been infiltrated and successfully attacked[3] and

[1]Davis (2007).

[2]Bundesamt für Sicherheit in der Informationstechnik (2014), p. 31.

[3]Langner (2013).

S. Meier (✉)
Universität Regensburg, Universitätsstraße 31, 93053 Regensburg, Germany
e-mail: stefan.meier@ur.de

W. Heintschel von Heinegg et al. (eds.), *Dehumanization of Warfare*,
DOI 10.1007/978-3-319-67266-3_8

even information security companies were not able to withstand these targeted attacks.[4,5]

To uncover the impact of digital attacks, the accrued digital evidence needs to be investigated. Therefore, methods from the digital forensic research area are used. But, attribution, as one of the central investigative questions (*Who did it?*)[6] cannot be easily answered by digital forensic methodologies. In theory, a perfect digital crime, i.e. a digital crime where any digital trace is wiped out is imaginable.[7] Even worse, a digital attacker can put fraudulent evidence in place to set a digital forensic investigation on the wrong track. However, through the complexity of todays' information systems, it is very hard to accomplish a perfect crime.[8,9] Errors committed by the attacker can quickly disclose fraudulent digital evidence.[10] However, a lot of myth exists in and around digital forensics and about the available capabilities. For that reason we show what digital forensics is, demonstrate the capabilities of digital forensics and uncover how the methodologies work in comparison to other forensic sciences.

The paper is structured as follows: In the next Sect. 2 the basics of digital forensics are lined out and it is shown how digital forensics works on an abstract level. Furthermore, the properties of digital evidence and its major characteristics are presented. In the subsequent Sect. 3 two example attack scenarios are sketched out and the emerging digital evidence and the potential outcomes of a digital forensic investigation are demonstrated. As examples the Stuxnet attack, outlined in Sect. 3.1, and the Distributed denial of service (DDoS) attack against the German Bundestag's Website, outlined in Sect. 3.2, have been chosen. In Sect. 4 we finally give a conclusion and a short outlook on the future development of digital forensics.

2 Digital Forensics

To get an understanding of digital forensics we first want to give a definition. While there exist many definitions for digital forensics,[11] we see digital forensics as a

---

[4]Coviello (2011).

[5]Kaspersky (2015).

[6]Rid and Buchanan (2014).

[7]Böhme et al. (2009).

[8]Böhme et al. (2009).

[9]Casey (2011), p. 26.

[10]Rid and Buchanan (2014).

[11]Slay et al. (2009).

scientific discipline in line with other non-digital forensic disciplines like serology, toxicology or ballistics. Therefore we define digital forensics as follows:

Definition 1 Digital forensics is defined as the application of computer science methodologies to solve questions of the legal system. Digital forensics especially provides methods to collect and analyze digital evidence for use in a court of law.[12,13]

While the definition clearly indicates the scientific domain of digital forensics, there is also another domain within the digital forensics research area. That domain deals with practical investigative questions like how to correctly obtain digital evidence rather than the central investigative questions: *who*, *what*, *where*, *why*, *when*, and *how*.[14] The two domains can be distinguished as follows:

1. The first domain which shares a common basic theory with other forensic sciences and comprises methods that deal with associations.[15]
2. The second domain which covers the entire digital investigative process and contains methods e.g. to search and recover digital evidence or to handle large volumes of data.[16]

In this paper we line out the capabilities and methods from the scientific domain of digital forensics and highlight a few practical implications resulting from the nature of digital evidence. In the next subsection we now give a short introduction to the common scientific principles and their utilization in digital forensic investigations.

2.1 Forensic Principles

Most of the established forensic concepts have evolved over the last decades.[17] Inman and Rudin[18] discussed and extended the historically evolved forensic science principles. Figure 1 gives an overview of the extended forensic principles. The crime event itself is characterized through the four overlapping domains *scene*, *witnesses*, *suspect* and *victim* and defines the border between the generation of evidence and the practice of forensic science.[19] The central processes for the

[12]Dewald and Freiling (2011), p. 49.

[13]Dewald (2012), p. 59.

[14]Inman and Rudin (2002).

[15]Dewald and Freiling (2012).

[16]Dewald and Freiling (2012).

[17]Inman and Rudin (2002).

[18]Inman and Rudin (2002).

[19]Inman and Rudin (2002).

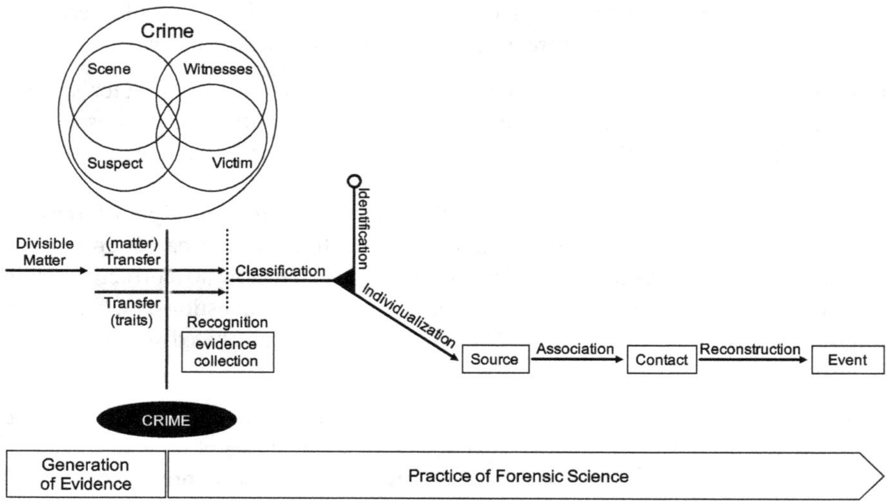

Fig. 1 Forensic paradigms (adapted from Inman and Rudin (2002))

practice of forensic science are identification, individualization, association, and reconstruction.[20]

The generation of evidence is explained by the principles of divisible matter in conjunction with the transfer of matter and the principle of the transfer of traits. For digital forensics, the concepts explaining the generation of evidence are limited to the transfer of traits.[21,22] However, the processes for the practice of forensic science can be applied without modification. Based on Inman and Rudin (2002) and Dewald and Freiling (2012) the process steps are defined as follows:

Identification In the identification phase potential evidence is recognized and classified. This step answers the investigative question: *what is it?*[23] Identification can be an end in itself, e.g. when illegal drugs or illegal pornography images are recognized, the legal system is not necessarily concerned with the marijuana field from which the drugs originated or the camera the pictures have been taken with.[24]

Individualization When identification is not an end in itself, the subsequent step is individualization where individual characteristics are determined in comparison to at least one additional item. The individualizations step

[20]Inman and Rudin (2002).

[21]Dewald (2012), p. 48f.

[22]Dardick et al. (2014), p. 156.

[23]Inman and Rudin (2002).

[24]Inman and Rudin (2002), Dewald and Freiling (2012).

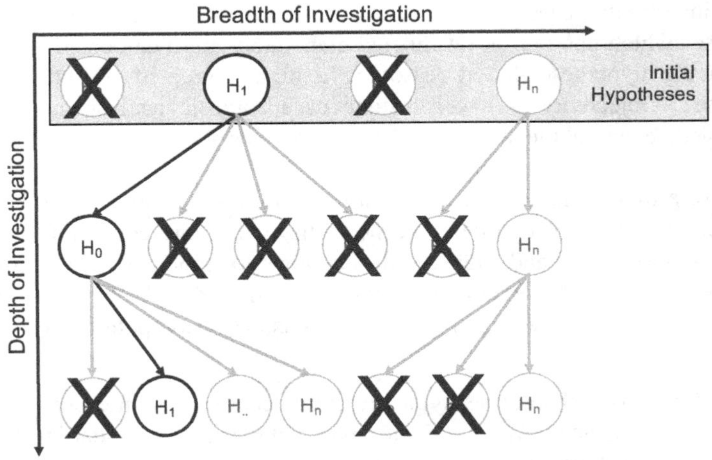

Fig. 2 Breadth and depth of the investigation based on hypotheses testing (adapted from Dewald (2012), p. 13)

asks if the items share a common origin and answers the questions: *which one is it?* or *whose is it?*[25]

Association Within the association process the contact between the source of the evidence and a target is inferred. The association process therefore evaluates and compares competing hypotheses regarding the origin of the two or more compared items. The inference is based on the outcomes of the identification and individualization steps.

Reconstruction Reconstruction is considered as the process to order associations in space and time. This step answers the questions: *where*, *how*, and *when*.[26]

As the association step indicated: Hypotheses testing is a crucial step to get an objective answer on whether a contact between two objects can be inferred. Hypothesis tests are essential for the scientific methodology in general.[27] With hypotheses tests, scientists examine if a proposed general principle can be inferred or if the hypotheses test invalidates the proposed principle.[28]

Hypotheses testing also implies that the validity of an association can never be verified.[29] Figure 2 shows how hypotheses tests can be used in forensic investigations. Starting with initial hypotheses about the circumstances of a crime, the investigator tries to invalidate each of the hypotheses. By refining the hypothe-

[25]Inman and Rudin (2002).

[26]Inman and Rudin (2002).

[27]Dewald (2012), p. 13.

[28]Inman and Rudin (2000), p. 5f.

[29]Inman and Rudin (2000), p. 5f.

ses the investigator tries to refute the hypotheses on a fine grained level. The hypotheses which can not be invalidated elaborately are then considered likely.[30] Through the hypotheses-based reasoning a high degree of objectivity can be maintained. Objectivity is a cornerstone of a forensic analysis and needs to be retained, even in the event of external pressures to reach specific conclusions.[31]

To give a more concrete picture on how the forensic principles can be applied in digital forensic investigations, we now show how one could infer a contact between computer A and website B, i.e. how to conclude that computer A visited website B. The example is based on the example from Dewald and Freiling (2012). The investigative hypothesis is: H_0: Computer A visited website B.

Identification A web browser typically stores copies of visited websites in the browser cache. An investigator may find files in the standard web browser cache folder on the hard drive of computer A. Furthermore he classifies these files as browser cache files of the installed browser.

Individualization After classifying the files as browser cache files, an investigator typically uses a tool to analyze the content of the files to determine which content has been cached. Furthermore, specific file characteristics like the creation date, size and content are determined.

Association Having enough individual characteristics like HTML[32] files, media files like images or videos and evidence from the files' content like the user name or people on the images only available from website B, the investigator is able to infer a contact between computer A and the specific website B and establishes this association.

2.2 Digital Evidence

After demonstrating how the common principles of forensic science can be applied to digital forensic investigations, the properties of digital evidence are shown. The web browser cache files mentioned as evidence in the example in the previous subsection represent only one concrete example of digital evidence. An e-mail message, a social network entry, a network packet, a SMS message on a mobile phone, your car's error log in the on board diagnose unit, etc. all these examples are

[30]Dewald (2012), p. 13f.

[31]Casey (2011), p. 24.

[32]HTML is the markup language used to semantically describe the structure of a website. HTML files and all referenced files like images are downloaded, rendered and presented by browsers like Firefox, Chrome, Internet Explorer or Edge.

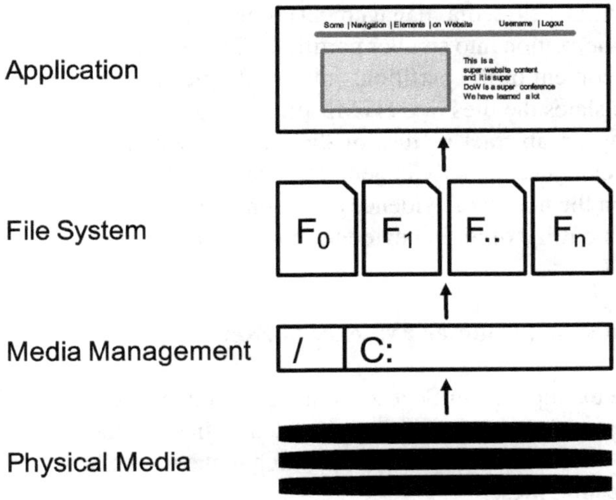

Fig. 3 Abstraction layers and layers of an HTML file (based on Carrier (2003))

concrete examples of digital evidence. Following the definition from Casey (2011), digital evidence is defined as follows:

Definition 2 "Digital evidence is defined as any data stored or transmitted using a computer that support or refute a theory of how an offense occurred or that address critical elements of the offense such as intent or alibi."[33]

Besides covering all examples of digital evidence above, the definition also includes an important aspect of evidence: Evidence is always related to a criminal event. This aspect also exists for physical evidence. Matter is always dividing, but only when the transfer of matter or traits occurs in conjunction with a criminal event the divided matter or the transferred traits become evidence.[34]

In contrast to many types of physical evidence like shoe prints or blood trails, digital evidence can not be reasonable interpreted directly by humans in the most cases. Although digital evidence is also physical evidence in the first place, e.g. in the form of magnetic waves on a cable or the magnetization of a hard disc surface, several abstraction levels are commonly needed.[35,36] Figure 3 shows the abstraction layers needed to interpret the aforementioned website content. The lowest (physical evidence) layer is represented via the physical hard disc containing the magnetized surface which is translated to a general understandable format for the next layer.

[33]Casey (2011), p. 7.

[34]Inman and Rudin (2002).

[35]Carrier (2003).

[36]Dewald and Freiling (2011), p. 36.

The subsequent layer (media management) translates the general representation of the disc's magnetization into smaller partitions. The following file system layer then translates the content of the partitions into single files and the application layer on top finally translates the files into HTML files.[37]

After giving an abstract picture of digital evidence and associations, we now describe how evidence actually is generated through the events of a system. In the next subsection the notion of evidence and counter evidence of events is introduced, followed by an outline of digital evidence characteristics in Sects. 2.2.2–2.2.6.

2.2.1 Evidence and Counter Evidence of Events

In the website example from Sect. 2.1 only a subsequent set of files from the hard disc of computer A is identified as digital evidence. In this concrete case, the browser cache files as well as the additional files which make up the browser program have been the interesting files.

To identify these files and in order to test the hypothesis, the investigator needs knowledge about how the system works and which events in detail potentially leave digital evidence. In the website case, the investigator assumed that a popular web browser was used to load the website. There are also other ways to load a website like a custom browser or a more simple command line tool that just downloads the main HTML file. These options can be covered through alternative hypotheses.

To test the aforementioned H_0 hypothesis, the investigator classified the found files as the one relating to the installed browser. The further individualization step was made through the identification of the unique website content. In the website case the characteristic HTML structure and the additional content like images, other related files and the user name are found to be characteristic evidence for the event of visiting website B with the specific browser on computer A. Therefore, the association was based on the reconstruction of events very likely run by computer A during the visit of website B. In addition to the definition of digital evidence above, we define the concrete evidence resulting from an event in a system as:

Definition 3 The evidence of a concrete event is the digital evidence, that is produced when the event is executed in the system.[38]

As some events in the system might produce the same subset of digital evidence, there is also the notion of characteristic evidence:

Definition 4 Characteristic evidence is the portion of digital evidence, which is unique for an event in a system in comparison with any other events which are possible in the system.[39]

---

[37]Carrier (2003).

[38]Dewald (2012), p. 83.

[39]Dewald (2012), p. 86.

In the website example the change of file access timestamps is characteristic for the usage of the popular web browser for the event of visiting website B. However, there are also other events which also change the same file access timestamps. The characteristic evidence is the digital evidence which has been saved in the browser cache, e.g. the unique HTML files and images from the specific website. So these evidence is characteristic evidence for the event of visiting website B.

In contrast to evidence which supports the hypothesis that an event occurred, there is also counter evidence supporting the hypothesis that the event did not occur. When the files in the website example are found to be altered in another way as by the event of visiting website B, e.g. by rewriting the files and changing the files' last modification timestamps instead of changing the files' access timestamp and writing the website's content into the cache files, this would be counter evidence.

Definition 5 Counter evidence is the portion of digital evidence, which is related to the same set of evidence of the event but alters this evidence on another way than the event in question does.[40]

By knowing which evidence is produced by which event in the system, a forensic investigator is able to infer on previous states of the system and reconstruct the detailed events which occurred by committing the crime. Starting with the next subsection we give an outline of general digital evidence characteristics which potentially effect the outcome of forensic investigations.

2.2.2 Evidence Volatility

In analogy to physical evidence, digital evidence has several levels of volatility. The three levels are: persistent, semi-persistent and volatile digital evidence[41,42]:

Persistent evidence Persistent evidence is non-volatile even without a current power supply to the system. Examples are: Data on hard discs, data on optical media like Blu-rays, CDs or DVDs, or data on flash media like USB flash drives or solid-state drives.

Semi-persistent evidence Semi-persistent evidence is non-volatile as long as the system has a current power supply. A typical example for this category is the computer memory (RAM).

Volatile evidence Volatile evidence is the data which is only temporary available on a running system with an active power supply. Examples are CPU caches or data on network cables.

The level of volatility especially raises questions for the practical collection of digital evidence. However, also the analysis phase can be affected, if volatile

[40]Dewald (2012), p. 91.

[41]Association of Chief Police Officers (2007).

[42]Dewald (2012), p. 39f.

evidence was not preserved and prior events can not be reconstructed due to missing digital evidence.

2.2.3 Evidence Manipulability

In common, manipulation or destruction of evidence can be detected very easily for physical evidence, as transfer occurs through every contact, i.e. also the event of manipulation leads to a division of matter and a transfer of matter or traits. Manipulation or the deletion of digital evidence is harder to detect, especially if evidence is not tamper-resistant, as digital evidence typically is.[43,44] As noted earlier, a perfect digital crime is imaginable in theory.[45] So, the opportunity to manipulate digital evidence easily and without leaving traces sets high requirements on the collection, preservation and handling of digital evidence. The determination of the authenticity and the preservation of the integrity of digital evidence are therefore crucial for the usefulness of digital evidence.

2.2.4 Authenticity

The authenticity of evidence is determined by two steps. First of all, it has to be determined that the evidence is what it claims to be, e.g. the cache files of the website example are analyzed and it is determined if these files are actually files resulting from the installed browser version. The second step determines the evidence's probative value.[46]

An important aspect of the authentication aspect is the maintenance and documentation of a chain of custody of evidence. This is especially important as digital evidence can be manipulated or destroyed very easy as explained in the previous subsection.[47]

2.2.5 Integrity

Besides authenticity, also integrity is a crucial aspect which needs to be preserved starting with the collection of a piece of evidence. To be able to later verify the integrity of digital evidence, a digital fingerprint is generated and secured against tampering during the evidence collection step.[48] As mentioned in Sect. 2.2.3, the

[43] Casey (2011), p. 26.
[44] Dewald (2012), p. 41.
[45] Böhme et al. (2009).
[46] Casey (2011), p. 21.
[47] Casey (2011), p. 21f.
[48] Casey (2011), p. 22.

Table 1 Levels of certainty in digital evidence (Casey 2011, p. 70)

Certainty level	Description/indicators	Commensurate qualification
C0	Evidence contradicts known facts	Erroneous/incorrect
C1	Evidence is highly questionable	Highly uncertain
C2	Only one source of evidence is not protected against tampering	Somewhat uncertain
C3	The source(s) of evidence are more difficult to tamper with but there is not enough evidence to support a firm conclusion or there are unexplained inconsistencies in the available evidence	Possible
C4	(a) Evidence is protected against tampering or (b) evidence is not protected against tampering but multiple, independent sources of evidence agree	Probable
C5	Agreement of evidence from multiple, independent sources that are protected against tampering. However, small uncertainties exist (e.g., temporal error and data loss)	Almost certain
C6	The evidence is tamperproof or has a high statistical confidence	Certain

importance of this protection of digital evidence against altering is a resulting requirement from the easy manipulability of digital evidence.

2.2.6 Levels of Certainty in Digital Evidence

After collecting and analyzing digital evidence and the formulation and testing of alternative hypotheses, the probability of error has to be determined to further evaluate the probative value of the collected evidence and to infer contact between two objects using the most likely hypothesis. Casey[49] defined seven levels of certainty to determine the certainty of digital evidence on a qualitative scale. The seven levels are outlined in Table 1.

2.3 Attribution with Digital Evidence

In the previous subsections, the characteristics of digital evidence as well as the methodology and scientific nature of digital forensics have been outlined. From the website example, it is obvious that digital forensics works like any other forensic discipline. Event reconstruction is based on hypotheses testing and therefore digital

[49]Casey (2011), p. 70.

investigations give an objective answer about which events likely happened on or in an information system in question. However, digital evidence is usually circumstantial, which makes it difficult to attribute a system event to an individual.[50] Even with good digital evidence in place, e.g. through a user name found in the contents of website B on computer A in the aforementioned website example, attribution solely based on digital evidence is difficult, because authentication mechanisms, if any, can be bypassed.[51]

In the next section, we demonstrate what information can be revealed by digital forensic investigations on two practice-driven examples. Furthermore, we show which conclusions can be inferred by these investigations.

3 Digital Forensics and Cyber Attacks

After the explanation of the basic concepts behind digital forensics, we now demonstrate what kind of digital evidence exists for the Stuxnet malware attack and the DDoS attack against the German Bundestag's website and what conclusions can be drawn from digital evidence. Both of the attacks are described on an abstract level. Additionally, the potential digital evidence is analyzed and the relevant conclusions are drawn. The Stuxnet attack has been chosen as it is one of the most sophisticated and best analyzed known malware attack. The DDoS attack against the German Bundestag's Website has been chosen as the attackers used a common attack pattern to seize a state target. However, one could easily pick another example as DDoS attacks are constantly executed against various targets.

3.1 Malware Attack

Malware[52] attacks are constantly happening. However, these broad attacks are nowadays mostly executed by organized crime and aim at peoples' money. Common attack patterns to lure money are intercepting transferals, extorting people by e.g. encrypting their files or just deceiving people by blocking access to their computer and displaying misleading images.

The more interesting malware attacks are targeted attacks with highly sophisticated malware. The most prominent and also one of the best analyzed example of malware up until now is Stuxnet. The outline below and the findings are based on Langner (2013) and Zetter (2011). Figure 4 gives an overview of Stuxnet's components and the attacked rotor drives. On a high level view, the Stuxnet malware

---

[50] Casey (2011), p. 26.

[51] Casey (2011), p. 26.

[52] Malware is a general term for viruses, trojan horses, hoaxes, worms, etc.

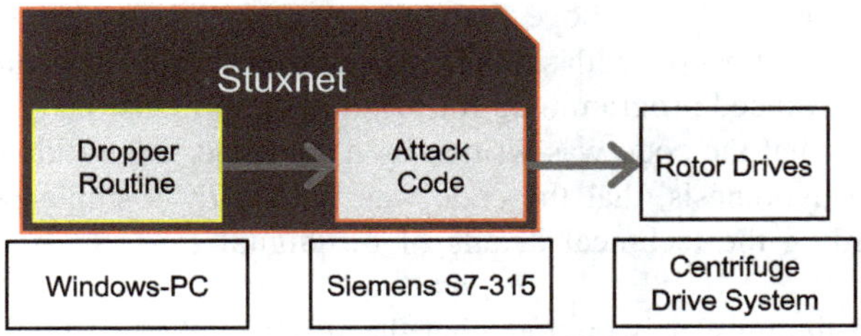

Fig. 4 Stuxnet components and deployment scheme (based on Langner (2013))

consists of a *Dropper Routine* and the actual *Attack Code*. The *Dropper Routine*'s purpose was to inject the *Attack Code* into Siemens S7-315 industrial control system components. These Siemens components are among others used to control rotor drives of uranium enrichment centrifuges, which have been the (physical) target of the Stuxnet malware attack. The second main purpose of the *Dropper Routine* was to deploy itself over private networks and USB flash drives. For this purpose the *Dropper Routine* had also a deployment mechanism.

As Stuxnet was initially detected by the public, researchers were amazed by this piece of software. The further forensic analysis of the Stuxnet program files, which are equal to the set of digital evidence, by several companies and research facilities revealed the full scale of Stuxnet's capabilities. The major findings are:

- There are various known versions of Stuxnet.
- The considered later Stuxnet versions carried two attack routines:
  - A routine to attack the pressure control (also part of the first version).
  - A routine to directly attack the rotor speed by manipulating the rotor drives.
- To deploy itself, Stuxnet used zero-day exploits for the Microsoft Windows operating system in addition with stolen digital certificates from driver vendors. The deployment mechanism was only found in the later Stuxnet versions.
- Stuxnet only tried to deploy and execute the *Attack Code* when it found a proper environment.
- The Stuxnet software is of high quality and very sophisticated, both in terms of attack and deployment routines.
- After infecting a new PC, Stuxnet provided information about the infected PC to one of two locations in the Internet.[53] These also called command-and-control servers[54] additionally let the attackers update Stuxnet.
- Every found Stuxnet sample contained information about the previous PC it came from.

[53]Stuxnet picked either *www.mypremierfutbol.com.* or *www.todaysfutbol.com* to report the information about the recently infected PC. The domains have been hosted in Malaysia and Denmark.

[54]A command-and-control server is common to operate a bunch of infected PCs. This architecture (server+infected PCs) is also the standard for so called botnets, which are often used to deploy non-targeted malware or to execute DDoS attacks.

Putting all pieces of the above findings together, the attack code was likely written by a group of people with skills in information security, uranium enrichment technologies, advanced programming for Windows and industrial control systems. The probability that the code was written by a so called script kiddie is very low. Therefore, the hypothesis, that the code was written by a single person is very unlikely regarding the technical details of the digital evidence, i.e. the Stuxnet program files.

From these findings, it is obvious that the script kiddie can be excluded from the group of suspects. However, all other groups remain likely to be suspect based on digital evidence as the *why* can never be addressed by a consideration of digital evidence.[55] However, digital evidence itself answered the questions *what* Stuxnet actually did, *how* it worked, *where* it was used, i.e. for which operating environment it was written for, and *when* it was created. So, with further information, the *where* has been narrowed down to the uranium enrichment facilities in Iran and the *when* and *who* have also been circumscribed.

3.2 Distributed Denial of Service Attack

Besides malware attacks, DDoS attacks are common against civilian and military targets. The direct effect of a successful attack is the unavailability of the target. In the case of an Internet service, the service is unavailable and therefore unusable for ordinary users. For civilian targets, a DDoS has mostly monetary effects, i.e. unavailability costs money for the service provider as well as for the users. For state or military targets as well as for critical infrastructure, the unavailability of a service can cost peoples' lives either directly or indirectly and it can make a state or military entity incapable.

In the following example we concentrate on a simple and common DDoS pattern. The case is based upon the DDoS attack against the German Bundestag website www.bundestag.de.[56] As a common attack pattern was used, this case explains also many other DDoS attacks.

Figure 5 visualizes the Internet with nodes (circles) and edges (lines). The nodes represent computing devices like webservers, personal computers, smartphones, Internet of Things devices, etc. The edges are the network connections between these nodes. The thickness of an edge visualizes the bandwidth or capacity of the connection. More capacity means a thicker line, like an interstate with additional lanes. Nodes itself have also a capacity, e.g. one node may serve a few thousand users concurrently (depending on its hardware and network design). The www.bundestag.de website is hosted, i.e. operated on one of the outer nodes with only one

[55] Inman and Rudin (2002).

[56] Kannenberg (2015).

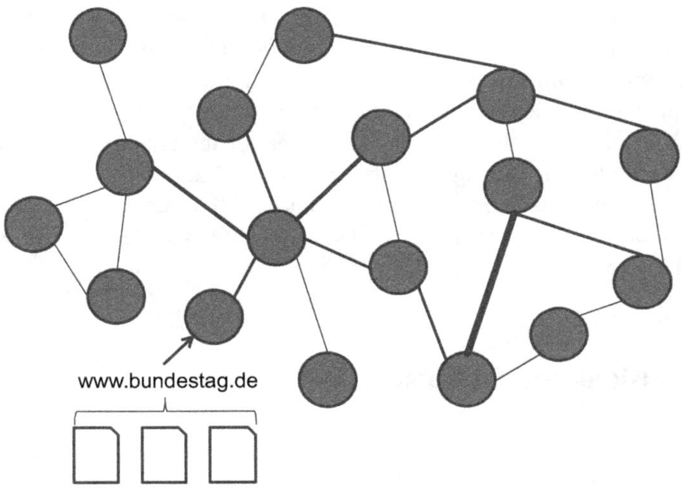

Fig. 5 The internet with the node hosting the German Bundestag's website

edge to one next node. As one can see, the Internet infrastructure is a distributed, non-centralized network of nodes.

The main target of a DDoS attack is the unavailability of the target by exceeding the number of concurrent users a node can serve at a time. So, if we have a node which can serve 1000 users at time, the 1001st user exceeds this limit and the node delivers an error for some users or is unable to react as its capacity is outrun. Considering Fig. 5, during the DDoS attack on the Bundestag website a lot of requests are sent by each other node to this one node hosting the Bundestag website. Through the distributed nature of the Internet, all requests take the best route available through the network of nodes and edges and finally hit the node which hosts the Bundestag's website. As the attack continues and more and more nodes constantly send requests, the node with the website at some point reaches its capacity limit and is therefore rendered unavailable.

By having an unavailable target, the DDoS attack has reached its goal. The impact of a simple DDoS attack can be uncovered easily, as the service is unavailable due to a high load on the affected node. So, to answer the investigative questions *what*, *where*, i.e. target of the attack, *how* and *when*, little digital evidence is needed. To uncover the *who*, digital forensics can only give a few hints as the origin of a DDoS attack can not be determined as easily. If the investigator can only gather digital evidence from the node hosting the website, he may find the IP addresses the requests came from. However the first problem arises as regular request to the website may not be distinguishable from the malicious requests. This is a common problem for DDoS attacks. So, with just the digital evidence from the node with the website, a digital forensic investigation can infer that a DDoS attack took place. Furthermore thousands of IP addresses may be found in the log files. However, the malicious attacker can not be inferred from these IP addresses, because IP addresses are likely to be manipulated and it is easy to use techniques to cover the own IP address. So, the IP addresses fall into the C0 or at best in the C1 level of certainty outlined in Sect. 2.2.6 and are therefore mostly useless.

If digital evidence can be collect from a higher number of nodes throughout the investigation, it is likely that e.g. the used botnet or the country the most requests came from are identified using the former useless IP addresses. However, even in this case, the *who* is just narrowed down to another Internet service like the botnet used. But today, even botnets are provided as a simple usable services to anybody who is willing to pay enough money.

4 Conclusions and Outlook

The two cases in the former Sect. 3 showed that attribution is hard to accomplish with solely digital evidence. Though digital evidence can solve a lot of important investigative questions. These answers give more information on the context and the circumstances of an attack and therefore are able to narrow down the group of suspects to a smaller group. Rid and Buchanan[57] also discussed the attribution problem. They found, that the technical level is not the only level needed to address when attributing a cyber attack. According to them, attribution is a multi-layered process run by organizations with a skilled and experienced team of investigators.

Besides the limitations of digital evidence regarding the attribution problem, digital forensics itself is still a very new forensic discipline. It evolves steadily and the capabilities become stronger from case to case. However, the available methods and the error rates for findings are not as established in digital forensics as in other forensic disciplines. But, digital forensics is developing and so are the skills, scientific proven methods and tools. As the two cases demonstrated, digital forensic methods can solve most of the investigative questions and can help to solve the remaining questions in the event of a cyber attack. However, there is a great deal of work ahead for the digital forensic research area, as the digital forensic methods need to keep up with new technological developments and the rising number of sophisticated digital attacks in the future.

References

Association of Chief Police Officers (2007) Good practice guide for computer-based electronic evidence. http://www.cps.gov.uk/legal/assets/uploads/files/ACPO_guidelines_computer_evidence%5B1%5D.pdf. Accessed 25 Aug 2015

Böhme R, Freiling FC, Gloe T, Kirchner M (2009) Multimedia-Forensik als Teildisziplin der digitalen Forensik. In: Informatik 2009: Im Focus das Leben, Beiträge der 39. Jahrestagung

[57] Rid and Buchanan (2014).

der Gesellschaft für Informatik e.V. (GI), Lübeck. Lecture Notes in Informatics (LNI), vol 154, pp 1537–1551

Bundesamt für Sicherheit in der Informationstechnik (2014) Die Lage der IT-Sicherheit in Deutschland 2014. https://www.bsi.bund.de/SharedDocs/Downloads/DE/BSI/Publikationen/Lageberichte/Lagebericht2014.pdf?__blob=publicationFile. Accessed 25 Aug 2015

Carrier B (2003) Defining digital forensic examination and analysis tools using abstraction layers. Int J Digit Evid 1(4):1–12

Casey E (2011) Digital evidence and computer crime: forensic science, computers, and the internet, 3rd edn. Academic, New York

Coviello A (2011) Open letter to RSA customers. http://www.sec.gov/Archives/edgar/data/790070/000119312511070159/dex991.htm. Accessed 30 July 2015

Dardick GS, Endicott-Popovsky B, Gladyshev P, Kemmerich T, Rudolph C (2014) Digital evidence and forensic readiness (Dagstuhl Seminar 14092). In: Dagstuhl Reports, vol 4(2) (2014), pp 150–190

Davis J (2007) Hackers take down the most wired country in Europe. http://archive.wired.com/politics/security/magazine/15-09/ff_estonia. Accessed 24 Aug 2015

Dewald A (2012) Formalisierung digitaler Spuren und ihre Einbettung in die Forensische Informatik. Dissertation, Universität Erlangen-Nürnberg

Dewald A, Freiling F (2011) Forensische informatik. Books on Demand, Norderstedt

Dewald A, Freiling F (2012) Is computer forensics a forensic science? In: Proceedings of current issues in IT security, Freiburg

Inman K, Rudin N (2000) Principles and practice of criminalistics: the profession of forensic science. CRC, Boca Raton

Inman K, Rudin N (2002) The origin of evidence. Forensic Sci Int 126:11–26

Kannenberg A (2015) DDoS-Attacke auf Web-Seiten von Kanzlerin und Bundestag. http://www.heise.de/newsticker/meldung/DDoS-Attacke-auf-Web-Seiten-von-Kanzlerin-und-Bundestag-2512871.html. Accessed 06 Aug 2015

Kaspersky E (2015) Kaspersky Lab investigates hacker attack on its own network. https://blog.kaspersky.com/kaspersky-statement-duqu-attack/. Accessed 30 July 2015

Langner R (2013) To kill a centrifuge: a technical analysis of what stuxnets creators tried to achieve. http://www.langner.com/en/wp-content/uploads/2013/11/To-kill-a-centrifuge.pdf. Accessed 25 Aug 2015

Rid T, Buchanan B (2014) Attributing cyber attacks. J Strateg Stud 38(1–2):4–37. https://doi.org/10.1080/01402390.2014.977382

Slay J, Lin YC, Turnbull B, Beckett J, Lin P (2009) Towards a formalization of digital forensics. In: Peterson G, Shenoi S (eds) Advances in digital forensics V. Fifth IFIP WG 11.9 International conference on digital forensics, Orlando, January 2009. IFIP advances in information and communication technology, vol 306. Springer, Berlin, Heidelberg, pp 37–47

Zetter K (2011) How digital detectives deciphered stuxnet, the most menacing malware in history. http://www.wired.com/2011/07/how-digital-detectives-deciphered-stuxnet/. Accessed 25 Aug 2015

CPU and Keyboard: Weapons of Mass Disruption?

Sigmar Stadlmeier

Abstract In a well-known paper on legal issues of computers and war Schmitt and others reminded their readers of the very basic fact that the "warriors" in computer wars would simply be "individuals armed with CPUs and keyboards". However, a few years later during a press conference in May 2009 U.S. President Barack Obama referred to instruments of cyber warfare as "weapons of mass disruption", obviously paraphrasing the well-known term of weapons of mass destruction (WMD) and having in mind the catastrophic consequences a coordinated attack on critical infrastructure could trigger. This paper primarily seeks to give an overview of the legal issues raised and investigate whether and to which extent cyber warfare operations fall in line with a trend of dehumanisation of warfare in the sense of reducing the human role or component in warfare. Given the limited size many basic legal questions of cyber warfare will be touched upon but not discussed in detail.

1 Introduction

In a well-known paper on legal issues of computers and war Schmitt and others reminded their readers of the very basic fact that the "warriors" in computer wars would simply be "individuals armed with CPUs and keyboards".[1] However, a few years later during a press conference in May 2009 U.S. President Barack Obama referred to instruments of cyber warfare as "weapons of mass disruption",[2] obviously paraphrasing the well-known term of weapons of mass destruction

S. Stadlmeier, Director, Institute of Public International Law, Air Law and International Relations, Johannes Kepler University Linz, Austria.

[1]Schmitt et al. (2004), p. 11.

[2]http://www.nytimes.com/2009/05/30/us/politics/30cyber.html Accessed 13 Dec 2016.

S. Stadlmeier (✉)
Institute of Public International Law, Air Law and International Relations, Johannes Kepler University Linz, Linz, Austria
e-mail: sigmar.stadlmeier@jku.at

W. Heintschel von Heinegg et al. (eds.), *Dehumanization of Warfare*,
DOI 10.1007/978-3-319-67266-3_9

(WMD) and having in mind the catastrophic consequences a coordinated attack on critical infrastructure could trigger. This paper primarily seeks to give an overview of the legal issues raised and investigate whether and to which extent cyber warfare operations fall in line with a trend of dehumanisation of warfare in the sense of reducing the human role or component in warfare. Given the limited size many basic legal questions of cyber warfare will be touched upon but not discussed in detail.

2 Vulnerability

A few months ago this writer was watching the screen while a routine update of a piece of software called Java Runtime Environment was being performed. As a clever piece of advertising Oracle Inc.,[3] the producer of Java, entertained the user waiting for the update to be completed that three *billion* devices worldwide use Java, the list including not only traditional computer-related devices but also cell phones, parking meters, ATMs, home security systems, TV sets and other devices—in short, almost anything which renders modern life so convenient and comfortable. At the same time, however, this highlights the substantial vulnerability of the information society and its "internet of things", given the fact that these things are all based on a few well-known IT platforms including their inherent risks. As long as those systems are running in a self-contained environment they must be attacked individually; as soon as they are connected to the internet malignant software code written for one of these platforms can no longer easily be contained within one sector or system but can—at least potentially—migrate to other sectors or systems ("springboard attack") because of the very fact that they are connected and share some basic platforms (suffice it to remind readers of the fact that the key transmission protocol of the net, TCP/IP,[4] is a dinosaur from the Seventies developed for free exchange—and not primarily for secure exchange—of information). The "internet of things" literally requires even critical ICS and SCADA systems[5] at least to have access points to the internet and increases the vulnerability to coordinated attacks on several systems of critical infrastructure, a term which typically includes IT, power supply, water supply, transport, financial markets, emergency services and the like. Even without such connection, however, clever attacks on such systems may trigger "cascaded failures" in the sense of one system failing because of the collapse of another (imagine a protracted blackout of power supply affecting the operation of water supply, fuel supply, transport and other systems depending on electric power). The overall effects of such a well-

[3]http://www.oracle.com/index.html Accessed 13 Dec 2016.

[4]Transmission Control Protocol/Internet Protocol.

[5]Industrial Control Software/Supervisory Control and Data Acquisition.

planned attack may shut down public life as we know it, and in this respect indeed compare to the secondary effects of WMD.

3 Methods and Means

3.1 Human Factor

Successful attacks will normally be based on intelligence and reconnaissance work, studying and investigating the strengths, weaknesses and particular vulnerabilities of a target, potential avenues of approach etc. In kinetic warfare[6] ruses of war typically aim at the human decision-making process, confusing and fooling the human mind by camouflage and credible but false information. Cyber warfare is no different in this respect, and in the cyber environment the exploitation of human weaknesses for the purpose of gaining access to systems is quite common. A very basic method is called *social engineering,* which in the IT context means extracting information and credentials (such as passwords, keycards or similar devices) by building up and subsequently exploiting social relationships, usually by pretending to be a colleague/supplier/customer/network administrator or the like. *Phishing* (password fishing) e-mails have become well-known even among average end-users but still do occur, especially in a more sophisticated and targeted version of *spear phishing,* approaching the targeted person under the faked identity of another person known to be well-known to and occasionally in contact with the targeted person. Statistics on white-collar crime indicate that up to 75% of illegal flow of information result from activities of insiders ("moles")[7]; moreover, attacks on self-contained systems not directly connected to (or at least not easily accessible from) the internet usually must compromise one or more insiders, e.g. by having him or her innocently use a manipulated data device such as a USB stick (there are speculations that the famous Stuxnet worm may thus have initially been inserted into the affected ICS system).[8] Recent trends such as BYOD *("bring your own device"),* however attractive in terms of saving some IT hardware costs for employers, open up whole new avenues of approach waiting to be exploited.

[6]In this paper the term *kinetic* (meaning the projection or application of kinetic energy) is used to distinguish traditional forms of warfare from cyber warfare.

[7]http://www.security-insider.de/themenbereiche/bedrohungen/menschliche-faktoren/articles/10122 9/ (Accessed 13 Dec 2016; source in German).

[8]http://www.businessinsider.com/stuxnet-was-far-more-dangerous-than-previous-thought-2013-11 ?IR=T Accessed 13 Dec 2016.

3.2 Machines

On the other hand, cyber warfare does introduce elements of "machine warfare" with very little human involvement. The classic *denial of service (DOS) attack,* bombarding a computer system with so many innocent requests for something that it is overloaded, hangs and therefore denies its normal service (hence the name) works best (especially against powerful machines) if the requests come from many different (remote) sources at almost the same time, "distributing" the attack workload (hence the name *distributed [remote] denial of service attack, D[R]DOS*). The most convenient platform for such an attack is a so-called *BotNet* (ro*bot*ic *net*work) in which the attacker (called the *Botmaster*) has gained remote control over a number of computers by inserting malicious code exploiting software vulnerabilities or other weaknesses (e.g. remote maintenance options), subsequently using these computers (called *zombies*) to launch an attack at a predetermined time on a predetermined target. Obviously large arrays of computers in industrial or business environments which run 24/7 are most suitable as potential zombies since they can be called to action at any time against targets several time zones away. In view of the amount of predetermination (through pre-programming by the human botmaster) a botnet would seem to be an automatic, rather than an autonomous, system, but within this limit it does indeed function with very little human input; especially the owners and/or users of the zombie computers may not even be aware of the fact that their computers have become an active contributor to a botnet.

4 Examples

Fortunately the world has not yet seen a genuine cyber war, but at this point in time examples for cyber operations can be demonstrated which could be—or indeed have been—used in armed conflict: In 2007 a number of DDOS attacks on computer systems serving mainly the media and financial sectors in Estonia caused major inconveniences and shut-downs for several days.[9] In the same year an Israeli air strike on Syrian targets may have been supported by a cyber operation affecting the Syrian radar surveillance system.[10] In 2008 during a hot phase of the conflict in Georgia a well-known Russian website openly invited attacks on 38 websites in Georgia; 13 sites were actually attacked. The latter events illustrate the most likely scenario to be expected in the near future, cyber operations as a part of comprehensive military operations in the course of an armed conflict. In 2010 *Stuxnet* was discovered as a very sophisticated custom-made worm intended

[9]Watts (2010), p. 397.

[10]Hathaway et al. (2012), p. 838.

to inflict particular versions of Siemens-made ICS used in Iranian facilities to control machines for uranium enrichment (*Stuxnet* manipulated the rate of rotation, rendered the output unusable and ultimately destroyed the machines).[11] *Stuxnet,* being an electronic worm, was capable of fully automatic operation and self-reproduction; on the other hand, the final example given here fully relies on the human factor again: In 2013, South Korean "patriotic hackers" responded to attacks against South Korean IT infrastructure (again media and financial sectors were targeted) originating from North Korea with a website defacement attack against the official North Korean government website (at that time actually hosted in China) by putting on to it a portrait of North Korean president Kim Jong Un, dressed in what looked like an ancient Roman toga, showing the president with a pig's nose, pig's ears and a mickey mouse tattoo on his chest, obviously to ridicule the North Korean president.

5 Legal Challenges

5.1 Structure of Cyber Conflict

The structural characteristics of cyber conflict *per se* challenge classic international law of the post-Westphalian Peace-period: Cyber operations transcend traditional archetypes of international law such as territory and boundaries; physical space is largely replaced or supplemented by virtual space (the attribution problem will be discussed later on). Cyber operations are well-suited for asymmetric conflicts: Hardware is relatively cheap and readily available off-the-shelf, and so is software (basic hacker tools are frequently available for free, more powerful software still comes at an affordable price). The basic equipment, both hardware and software, is classic *dual use* material: Computers can be used for all sorts of things, and source code for either a mouse driver or a virus can be written in any standard programming language. Hence, as opposed to traditional kinetic *war material,* the proliferation of equipment cannot be controlled; equipment is therefore easily and lawfully available to governments and individuals alike. Unlawful cyber activities may appear as plain crime, organised crime, terrorism or warfare, and more often than not nation States will be facing non-State actors, together with the challenge of identifying the proper legal domain to deal with such actors.

[11] http://www.businessinsider.com/stuxnet-was-far-more-dangerous-than-previous-thought-2013-11?IR=T Accessed 13 Dec 2016.

5.2 *Selected Issues of Ius Ad Bellum*

5.2.1 Force, Aggression, Attack?

For cyber operations to become relevant under the law of armed conflict a certain intensity and overall coherent plan behind (among other things) are required. When would a cyber operation amount to an "attack", and which one of the four thresholds offered in international law should apply? Would cyber operations constitute any of the following, "violent acts against the adversary",[12] "use of force",[13] "act of aggression"[14] or "armed attack"[15]? For lack of any more specific guidance most authors rely on the *Nicaragua vs U.S.* case in which the ICJ held within the context of a subconventional kinetic conflict that only massive cases of "use of force" would because of their scale and effects amount to an "armed attack".[16] The question remains whether this is to be applied *ex ante,* which would bring the mere targeting of critical infrastructure or a comparable hostile intent within the notion of "attack", or whether actual effects amounting to those described by the ICJ must be demonstrated. The majority of writers has considered an effect-based approach as most appropriate to international law, and has understood the rudimentary guidance of the ICJ as requiring death or physical injury of people and/or the destruction of property, or massive medium to long-term disruption of critical infrastructure systems of a State equal to their physical destruction.[17] As far as the application of this concept to cyber operations is concerned the majority seems to agree with *Schmitt* who proposed a set of criteria,[18] namely first and foremost severity, together with immediacy, directness, invasiveness, measurability, presumptive legitimacy and responsibility (pointing once more to the attribution problem to be discussed separately) to determine whether a cyber operation indeed amounts to an attack by a State. Finally, the criterion of "armed" attack must not be (and has not been) understood in a narrow sense[19] but does include indirect uses of force and even improvised tools, as highlighted by NATO's reaction to the 09/11 events, declaring a *casus foederis* for the first time in history, referring to the North

[12]Art 49 para 1 Additional Protocol 1977 to the 1949 Geneva Conventions.

[13]Art 2 para 4 UN Charter.

[14]Art 39 UN Charter.

[15]Art 51 UN Charter.

[16]ICJ, Case Concerning Military and Paramilitary Activities in and against Nicaragua (Nicaragua v. United States of America), Merits, ICJ Rep. 1986, pp. 14–150, 118.

[17]See the overview in Ziolkowski (2012), p. 298.

[18]Schmitt (2011), p. 576; Schmitt (2013), Tallinn Manual, Rule 11. – For a critical review see Ziolkowski (2012), p. 301 et seq.

[19]See again Ziolkowski (2012), p. 298.

Atlantic Treaty and its provision on collective self-defence which in turn refers to Art 51 UN-Charter and the concept of collective defence against an armed attack.[20]

5.2.2 Target?

The Schmitt criteria mentioned above focus on the role of the attacker and—generally speaking—the scale and directness of the resulting consequences. In the view of this writer, both Schmitt and critical commentators focus too much on the role of the attacker of "Use of Force" and not enough on the target: According to Art 2 para 4 UN Charter "Use of Force" must be directed against certain fundamental values of nation States protected by international law, i.e. territorial integrity and political independence. One might by a combination of the "directness" and "invasiveness" identify this element; nevertheless it should be made clear that an invasion into these protected positions of a State under international law is the qualitative key (such as severity is the quantitative one) to the relevant thresholds of use of force amounting to armed attack. This point has been nicely made in the definition of a cyber attack as "any action taken to undermine the functions of a computer network for a political or national security purpose" as given by Hathaway et al.[21]

5.2.3 Attribution

Responsibility of a State (the last one of the Schmitt criteria quoted above) is the consequence of a violation of an international obligation attributable to a State.[22] Attribution is one of the most difficult issues (albeit in terms of facts rather than law) in the context of Cyber operations. For instance, how could one attribute "patriotic hackers", such as in the website defacement case involving the North Korean governmental website mentioned above? Again the ICJ has given some guidance in the abovementioned *Nicaragua* case by requiring effective control of a State over particular actions of non-State actors in order to trigger State responsibility for such action. The paramount problem is proof rather than definition, and there are many ways to camouflage one's identity in order to prevent attribution: The key to identification in the internet is an IP address unique to each user—or so it seems. An IP address, however, is at best unique to a machine, rather than any particular user; moreover, under older versions of TCP/IP a "unique" IP address may be allocated to a mere router; several machines connecting to the Internet via

[20] Art 5 North Atlantic Treaty; see http://www.nato.int/docu/pr/2001/p01-124e.htm Accessed 13 Dec 2016.

[21] Hathaway et al. (2012), p. 821.

[22] Art 2 ILC Draft Articles on State Responsibility.

this router all will appear under the router's IP address.[23] Moreover, IP addresses can be hidden, manipulated or disguised (e.g. by accessing the Internet though a proxy server), requiring additional forensic work in order to trace an attack back to a particular attacker (and remember: Some of the actors in the process may not even be aware of their being part of it, such as in the case of "zombie" computers in a botnet). The good news is that the "digital footprint" of an attack may tell a few things in terms of capabilities required to perform it, thus leaving only those having such capabilities: *Stuxnet* made simultaneous use of several "zero day exploits", i.e. newly found weaknesses of software, exploited right after discovery and before either the developer of the software or developers of protective software even learn about that particular weakness.[24] One such discovery may be luck; multiple discoveries point to systematic research. Moreover, anyone who designs a piece of malware such as Stuxnet would have to test whether it would work as designed on Siemens ICS which is nowhere near as common and widespread as, for instance, Oracle's Java or Microsoft Windows, and finally would have to test whether it indeed had the desired effects on hardware for uranium enrichment which is available under a tight control regime only. Development and testbed by far exceed the capabilities of a bunch of patriotic hackers acting on their own in a backyard and leave only wealthy nation States with access to the hardware and technology involved.

5.2.4 What If Not . . . ?

A cyber operation which fails to meet the scale criterion could still qualify as a "mere" unlawful intervention; the other problems, including the need to attribute the operation to a State, remain. Likewise, a reaction of the victim of the attack as a countermeasure requires the attribution of the violation to an offending State, since countermeasures may not harm third States.[25] Countermeasures of a State affected by a cyber operation which cannot be attributed to another State will probably harm legitimate interests of other States and will *prima facie* be unlawful unless covered by circumstances precluding wrongfulness under the ILC draft articles. Of those, distress and necessity might be helpful, if the countermeasure is the only reasonable means of saving human life or to safeguard an essential interest against a grave and imminent peril and does not seriously harm the interests of other States.[26] Both however cannot be invoked if the victim of a cyber operation contributed to the effects (e.g. by neglecting reasonable precautions against such operations).

---

[23]The most recent version IPv6 does identify each single machine but has not yet become universal standard.

[24]https://en.wikipedia.org/wiki/Zero-day_(computing) Accessed 13 Dec 2016.

[25]See Arts 49–53 of the ILC Draft Articles on State Responsibility.

[26]Arts 24–25 of the ILC Draft Articles on State Responsibility.

5.3 *Selected Issues of ius in bello*

The challenges presented by cyber warfare operations with regard to the *ius in bello* are not significantly different from those raised in the context of kinetic warfare and do not support any conclusions as to a *per se* dehumanising effect, as long as cyber operations remain under direct human control (directly executed or pre-programmed). This has been the case in all cyber operations observed so far (see above). Some elements even increase human involvement in an undesirable way, as will be demonstrated.

5.3.1 Legitimate Targets

In armed conflict, only military objects are legitimate targets. Attacks must be limited to those objects which by their nature, location, purpose or use make an effective contribution to military action and whose total or partial destruction, capture or neutralization, in the circumstances ruling at the time, offers a definite military advantage.[27] This is a dual criterion: The target must be militarily useful to party A, and its neutralisation must offer a definite advantage to party B. In a cyber context, destroying or neutralising random systems will not always offer a definite advantage (let us remember that the very idea of the internet is redundancy in routing information, rendering one single decapitation strike impossible); neutralising the domain name servers of a network however does indeed render all members of the affected network *incommunicado* until the damage is repaired: The domain name servers (DNS) can be regarded as the "phonebooks" of a network and resolve user addresses (typically user@host.domain) to IP addresses and vice versa. Without their service neither incoming nor outgoing messages can be routed. The effect of taking out the DNSs exactly equals jamming a whole radio net or blowing up a main network node.

5.3.2 Proportionality

The mere fact that objects do not only contribute to military action but serve other purposes too does not exclude them from being legitimate targets but merely requires a proportionality assessment, i.e. whether the expected collateral damage is acceptable in the light of the military advantage sought. The latter must in this case be "concrete and direct",[28] which is a stricter criterion than "definite". This makes sense since the target definition decides whether an object may be attacked at all; the proportionality review is meant to limit any such attack in terms of methods and

[27] Art 51 AP I 1977 to GC 1949.

[28] Art 57 AP I 1977 to GC 1949.

means to what is absolutely necessary to achieve a particular goal. For example, if military air surveillance serves military purposes only there will be no major proportionality problem when launching a cyber attack; things are different if it also serves civilian purposes (directly by handling civilian traffic or indirectly by contributing to a comprehensive "system image" for civilian air traffic control). In the latter case a cyber attack resulting in blank screens, resulting in a loss of guidance for all air traffic, would have to pass a rather strict proportionality test.

Such a test may, however, produce surprising results if one assesses "collateral damage" *inter alia* in terms of effort required to reverse it. Suppose a bridge is leading across a waterway, and traffic (both road and rail) uses both the waterway and the bridge. Certainly military operations rely on, and benefit from, civilian transport infrastructure; equally certainly civilian traffic will use such infrastructure. The military advantage of rendering impassable such a major node for three different means of transport is quite obvious, given the paramount importance of logistics for modern armies (a fact not properly appreciated by those unfamiliar with military operations). If our ficticious bridge is a fixed structure there will be no alternative to blowing it up, either by an attack using smart (e.g. laser-guided) bombs or a commando raid. If successful, all three traffic routes will be blocked, since the bridge will be gone for road and rail, and its collapsed remnants will block the shipping channel. After the conflict the remnants will have to be removed and the bridge rebuilt, both at great cost.[29] However, if our bridge can be raised and lowered (see Fig. 1,[30] a process which nowadays can be expected to be computer-controlled), the key question during a proportionality review will be "bomb or hack?". All three traffic roads can easily be blocked by jamming the bridge while half open (road and rail can no longer pass, and shipping cannot either for lack of height clearance), and the answer will be obvious if one considers the post-conflict scenario and the costs and efforts of "rebuild or re-boot?" Finally, a mere comparison of the operational (not to mention the political) costs of even a sophisticated piece of malware such as *Stuxnet* on the one hand and a kinetic attack on the enrichment facilities in question on the other hand illustrates the attractiveness of cyber means for this kind of operation.[31]

[29]This was the fate of the bridges across the Danube in Belgrade, destroyed by NATO in 1999 during *Operation Allied Force*.

[30]The bridge shown in Fig. 1 is a combined rail/road bridge in the old harbour of Victoria, British Columbia, Canada. The photo was taken in 2008 by this writer while the bridge was being raised to allow a ship to pass underneath.

[31]General Michael Hayden, a former director of the National Security Agency and CIA in the US, estimated that the worm cost an estimated US$1 million to create. http://www.thenational.ae/business/industry-insights/technology/former-cia-chief-speaks-out-on-iran-stuxnet-attack Accessed 13 Dec 2016.

Fig. 1 Johnson Street Bridge (2008), Victoria, British Columbia, Canada

5.3.3 Special Objects

Critical infrastructure includes power supply, and installations related to power supply are among the classic examples of "works and installations containing dangerous forces" within the meaning of Art 56 AP I 1977. Not only nuclear reactors but also dams of hydro power plants or the fuel storage facilities of steam-driven power generators contain forces, which can devastate large areas when released. Their protection under Art 56 is not absolute but ceases in case of regular, significant and direct support of military action, and if such attack is the only feasible way to terminate such support. Both for kinetic and cyber attacks, however, the latter issue will be crucial: If an attack on transformer stations and/or power lines could likewise terminate such support, an attack on the protected work or installation is prohibited. Cascaded failures, however (other vital systems failing due to the lack of power supply) will still occur.

5.3.4 Indiscriminate Attacks

Art 51 AP I 1977 requires the protection of the civilian population under all circumstances. To that end, the principle of distinction generally requires to distinguish at all times between legitimate targets (combatants and military objects) and protected persons or objects. Attacks which do not or cannot so distinguish because they are not or cannot be directed against a specific military objective, or the effects of which cannot be limited to such military objectives are considered indiscriminate attacks, and are prohibited. Some methods and means of cyber operations would potentially be indiscriminate within this definition: Worms, as opposed to viruses, can by definition self-reproduce and spread even without human intervention,[32] depending on their programming (e.g. sending themselves to every address in the e-mail address book of an infected system, repeating that process at every destination thus reached, resulting in an automatic—but not autonomous—chain reaction beyond the control of those who first released it). Such a worm cannot by definition be limited to legitimate targets, and its release would almost certainly constitute an indiscriminate attack. On the other hand and depending on the particular programming, even a worm such as *Stuxnet* can be particularly tailored to some distinctive systems of military importance.

5.3.5 Personnel

The rules of the law of armed conflict concerning personnel are largely influenced by a completely outdated conflict model of uniformed armies fighting each other in a symmetric conflict, supported by a limited number of services (medics, chaplains, supply contractors) but focusing on the combatant as the bearer of arms, clearly distinguishable from protected civilians. Only the 1977 AP I has departed from this model by acknowledging that in certain types of conflict such distinction cannot be made at all times, and stated as a minimum requirement that such distinction must at least be possible during each military engagement and during deployment preceding such engagement while visible to the enemy.[33] Generally, however, combatants have been supposed *inter alia* to bear their arms openly and have a fixed distinctive emblem recognizable at a distance ever since the famous Hague Regulations became effective.[34] It is difficult, to say the least, how these requirements could be observed in cyber warfare. Bearing arms openly is not easily achieved when no arms in the classic sense are being used at all; however, kinetic warfare has already departed from this requirement to a considerable degree: During engagements (e.g. in aerial warfare) beyond visual range arms cannot actually be seen but only assumed (even a distinctive radar profile consistent with a certain type of military aircraft does not reveal much of its actual payload), and even the identification of a target under such conditions remains difficult (a military tanker aircraft may have the same radar profile as its civilian airliner twin from which it derives). A distinctive emblem

[32] While viruses require some user action (e.g. opening of an e-mail attachment), worms do not.

[33] Art 44 AP 1 1977.

[34] Art 1 Hague Regulations on the Rules and Customs of War on Land = Annex to the IV Hague Convention 1907.

"recognizable at a distance" may perhaps be possible through the use of reserved military IP address space and/or dedicated *.mil domain names. The issue must be resolved soon, since many modern armies have already reacted to the technical development and formed cyber warfare units. E.g. both the U.S. and China have added "space" and "cyber" to the classic military domains of "land", "sea" and "air" and train military units for these future battlefields.[35]

The problems however are not limited to combatants in the classic sense: "Outsourcing" has become as popular in the military business as in any other business. Of all the services and capabilities required by a modern army very few are genuinely "military" by nature; much of it can be—and indeed is—supplied by commercial contractors, in particular logistics, transport, IT and maintenance of sophisticated equipment. In the context of armed conflict, this raises the issue of status of personnel involved in such services. The 1949 Geneva Conventions I and II use the term "supply contractors" for this personnel,[36] which will certainly be relevant in a cyber warfare context: Unlike subcontracting support tasks *en bloc* to private military and security companies (PSMCs), IT support in a cyber warfare context requires highly specialized and trained individuals working in close liaison with the military. Even if not fully integrated into the armed forces (which would make them combatants!)[37] they must be regarded as "following the armed forces" in the sense of the 1949 Geneva Conventions quoted above. On the one hand, the principles of distinction and protection of civilians require a separation of combatants from protected civilians, and military objects from protected civilian ones,[38] on the other hand the outsourcing of military services to civilian suppliers produces the opposite results: After capture, both combatants and civilian supply contractors are entitled to POW status and protection; while in action, combatants are legitimate targets, and supply contractors will end up as collateral damage, subject only to the proportionality test discussed above. Hence recent outsourcing trends tend to involve more human beings in cyber warfare than reduce human involvement.

6 Conclusion

While the technical aspects of cyber warfare might indeed suggest a reduction of the human dimension and role in warfare, a closer look reveals that this is limited to certain methods and means. Such reduction in some areas is offset by increased human involvement in other areas, creating additional problems under the law of armed conflict. Moreover, the effects of cyber warfare are not limited

[35] See e.g. https://www.stratcom.mil/factsheets/2/Cyber_Command/ Accessed 13 Dec 2016.

[36] Art 13 GC 1949 I and II.

[37] Art 43 para 1 AP I 1977.

[38] Art 58 AP I 1977.

to the machines involved but affect human society as a whole. Key issues of legal qualification, such as the notion of attack, attribution, and the adherence to traditional *ius in bello* rules are less significant with regard to the amount of human involvement but rather with regard to the role of humans and the challenges resulting from this role to the classic Westphalian system of international law.

References

Hathaway OA, Crootof R, Levitz P, Nix H, Nowlan A, Perdue W, Spiegel J (2012) The law of cyber-attack. Calif Law Rev 100:817

Schmitt MN, Harrison Dinniss H, Wingfield T (2004) Computers and war: the legal battlespace. Background paper prepared for informal high-level expert meeting on current challenges to international humanitarian law, Cambridge, 25–27 June 2004

Schmitt MN (2011) Cyber operations and the Ius ad Bellum revisited. Villanova Law Rev 56:576

Schmitt MN (ed) (2013) Tallinn manual on the international law applicable to cyber warfare. Cambridge University Press, Cambridge

Watts S (2010) Combatant status and computer network attack. Va J Int Law 50:392

Ziolkowski K (2012) Ius ad bellum in cyberspace – some thoughts on the "Schmitt Criteria" fort the use of force. In: Czosseck C, Ottis R, Ziolkowski K (eds) 4th international conference on cyber conflict, Tallinn. See https://ccdcoe.org/publications/2012proceedings/5_3_Ziolkowski_IusAdBellumInCyberspace.pdf

Part V
Specific Aspects of Dehumanization

Soldier 2.0: Military Human Enhancement and International Law

Heather A. Harrison Dinniss and Jann K. Kleffner

Abstract Advances in technologies that could endow humans with physical or mental abilities that go beyond the statistically normal level of functioning are occurring at an incredible pace. The use of these human enhancement technologies by the military, for instance in the spheres of biotechnology, cybernetics and prosthetics, raise a number of questions under the international legal frameworks governing military technology, namely the law of armed conflict and human rights law. The article examines these frameworks with a focus on weapons law, the law pertaining to the detention of and by 'enhanced individuals', the human rights of those individuals and their individual responsibility for the actions that they take while under the influence of enhancements.

1 Introduction

Whether it is science fiction classics such as Huxley's *Brave New World* or Heinlein's *Starship Troopers*, beloved characters like Captain America or the more modern versions shown in films such as 2012's *The Bourne Legacy*, the idea of human enhancement, particularly the enhancement of members of the armed forces, is one that has inspired, alarmed and enthralled generations of authors, readers,

Heather A. Harrison Dinniss: Senior Lecturer in International Law, Swedish Defence University.
Jann K. Kleffner: Professor of International Law and Head of the International Law Centre, Swedish Defence University.

This article previously appeared in Vol 92. ***International Law Studies (2016) 432–482***, and is reproduced with the kind permission of the U.S. Naval War College.

H.A. Harrison Dinniss (✉)
Centre for International and Operational Law, Swedish Defence University, Stockholm, Sweden
e-mail: heather.harrison-dinniss@fhs.se

J.K. Kleffner (✉)
Centre for International and Operational Law, Swedish Defence University, Stockholm, Sweden
e-mail: jann.kleffner@fhs.se

W. Heintschel von Heinegg et al. (eds.), *Dehumanization of Warfare*,
DOI 10.1007/978-3-319-67266-3_10

movie-goers and real-world scientists alike.[1] As twenty-first century science takes great strides into making these imagined worlds possible, armed forces and their legal advisors need to reflect carefully on the implications of these technologies for the international legal frameworks under which they operate. Of particular concern are the law of armed conflict and human rights law and their application vis-á-vis technologically enhanced armed force personnel and those who are subjected to actions by such personnel.

For the purpose of the present analysis, human enhancement is defined as the process of endowing an individual with an ability that goes beyond the typical level or statistically normal range of functioning for humans generally (or the personal unenhanced capabilities of a particular individual), where the ability is either integrated into the body or is so closely worn or connected that it confers an advantage similar to an internal or organic enhancement that transforms the person.[2] While the science is constantly and rapidly evolving, human enhancement technologies can be broadly divided into three main categories: biochemical, cybernetic (or brain-machine interfaces) and prosthetic.[3] Although there is a certain amount of overlap between the categories (particularly between cybernetics and prosthetics), each category raises slightly different legal issues.

Biochemical enhancement refers to the use of drugs or other pharmacological agents to enhance specific functions in the body. In recent years, one of the most talked about examples of biochemical enhancement is the use of modafinil. A neuro-enhancing drug originally designed for the treatment of narcolepsy, in otherwise healthy individuals modafinil acts as a cognitive-enhancer, as well as increasing wakefulness and the ability to focus. Not only have these properties made it the underground drug of choice of students taking examinations or writing essays,[4] but modafinil has also been approved for use by the armed forces of both the United States and France, and has been investigated by the United Kingdom and Canada (among others) for possible military use.[5] While the ability to operate effectively for long periods of time without the need for sleep has long been the holy grail of potential military enhancement, biochemical enhancement of any kind is not an automatic panacea. It raises complex legal issues in terms of the treatment of enhanced soldiers detained by an adversary (for example, whether sleep deprivation would still constitute inhuman and degrading treatment of personnel who have taken modafinil or a more permanent variation thereof) and an individual's ability to

[1] Huxley (1932), Heinlein (1959), and The Bourne Legacy (2012). Marvel Comic's Captain America first appeared in a comic series of the same name in March 1941. See also, Johnson (2012), pp. 34 ff.

[2] Lin (2010), p. 313; Daniels (2000), p. 309.

[3] One might also point to genetic enhancement or gene therapy as a further possible category of enhancement. However it has not been included in this research, other than as a passing reference, as it remains highly speculative.

[4] See Thomson (2015).

[5] Saletan (2013).

refuse to be subjected to the enhancement, as well as issues regarding the effect on individual accountability for any actions taken while under the influence of an enhancement.

The need for sleep is not the only "human failing" that armed forces are looking to manipulate with biochemical enhancement. Other drugs are being investigated for their utility in reducing the incidence of post-traumatic stress disorder and the reduction of fear—long implicated in the commission of war crimes—as well as reducing traumatic memory formation.[6]

The field of cybernetics has also provided many technological advances of interest to the military. Brain-machine interfaces, also known as neural interface systems, attempt to connect the brain directly to a machine without the need for manual input, such as a keyboard, joystick or other device. The interfaces use electrodes (whether surgically implanted in the brain or merely resting on the scalp) to record and translate the user's brain signals into commands that operate computer-controlled devices. The technology has been used both to actively control an external device (e.g., the ability to operate drones with the mind[7]) and to passively sift information by using the brain's power to unconsciously detect anomalies in large amounts of data.[8]

Prosthetic technology has made huge strides in the past 2 years. Scientists have developed prosthetics capable of providing sensory feedback from replacement limbs[9]; ways of linking prosthetics using neural interfaces or nerve endings to provide thought-controlled movement[10]; and visual prosthetics to provide augmented vision for those with normal vision and restore vision to those who have lost it.[11] Armed forces are already using exoskeletons such as Lockheed Martin's Human Universal Load Carrier and investigating prosthetics to enable soldiers to target their weapons more accurately.[12]

In this article, we will address these three forms of military human enhancement technologies from an international legal perspective by examining, first, questions that arise under the law of armed conflict and, second, questions that arise under international human rights law.

---

[6]Tennison and Moreno (2012).

[7]See e.g. Bienaimé (2014). In fact the technology has spread sufficiently enough that 2016 saw the first mind-controlled drone race organised by the University of Florida.

[8]See e.g. the Cognitive threat warning system (CT2WS) developed by the U.S. Army and DARPA which detects brainwaves in order to signal when the subconscious evaluates a visual threat: Ungerleider (2012).

[9]See Talbot (2013).

[10]Ibid.; Drummond (2012).

[11]See e.g. Sherwell (2015).

[12]Choi (2015); Black (2013), http://www.bloomberg.com/news/articles/2013-03-19/iron-man-meets-hulc-as-lockheed-enters-exoskeletons-race.

2 Law of Armed Conflict

2.1 *Weapons Review*

In the study, development, acquisition or adoption of a new weapon or new means and methods of warfare, a State which is a party to Additional Protocol I (AP I) is under an affirmative obligation to determine whether its employment would be prohibited under the Protocol or any other rule of international law applicable to that State.[13] A number of States not party to AP I, including the United States, have adopted review mechanisms along similar lines, albeit as a matter of policy rather than international legal obligation.[14] The first issue that arises in connection with the use of military human enhancement technologies is whether and to what extent such technologies trigger the requirement of a review as provided for in AP I and/or domestic regulations and practice. An answer to that query hinges in turn on the question of whether human enhancement technologies—or more precisely a given specific human enhancement technology—qualify as a weapon or means or method of warfare.

The term "weapon" is understood to refer to a means of warfare used in combat operations that is capable of causing either injury to or death of persons, or damage to or destruction of objects.[15] As such, a weapon "connotes an offensive capability that can be applied to a military object or enemy combatant."[16] The term "means of warfare" is broader than a weapon inasmuch as it "extends ... to weapon systems or platforms employed for the purposes of attack,"[17] and includes associated equipment used directly to deliver force during hostilities.[18] Means of warfare, and its sub-category of weapons, hence apply only in the context of attacks, that is "acts of violence against the adversary, whether in offence or in defence."[19] Finally, methods of warfare refers to activities designed to adversely affect the enemy's military operations or military capacity and extends to the various general categories of operations (bombing, ground, close-air support, etc.), as well as the

[13]Protocol Additional to the Geneva Conventions of 12 August 1949, and Relating to the Protection of Victims of International Armed Conflicts art. 36, June 8, 1977, 1125 U.N.T.S. 3 [hereinafter Additional Protocol I].

[14]U.S. Deputy Secretary of Defense (2003). See also Office of the General Counsel, U.S. Department of Defense (2015), paras. 6.2, 19.20.1.2. See also, Israel: Yaron (2016).

[15]See Program on humanitarian policy and conflict research (2009), pp. 1 ff.; Program on humanitarian policy and conflict research (2010), p. 55.

[16]McClelland (2003), p. 404. See also Boothby (2009), p. 4.

[17]Program on humanitarian policy and conflict research (2009), r. 1(t).

[18]Boothby (2009), p. 4.

[19]Additional Protocol I, supra note 13, art. 49(1).

specific tactics used for attack.[20] A common shorthand for methods of warfare is "the way in which [weapons] are used" in hostilities.[21]

Despite the occasional assertion to the contrary,[22] these working definitions suggest that the enhanced human soldier, per se, is not to be considered a weapon, because it is not the *person* that constitutes the offensive capability that can be applied to military objectives or enemy combatants. At the current state of development in the realm of human enhancement technologies, a distinction between the human, on the one hand, and the enhancement technology, on the other, remains possible since the use of the technology does not convert the human into an object that could be considered a weapon. While further developments may lead us down a path that makes that distinction more complicated, we do not seem to have arrived at the crossroads where enhancement technologies transmute humans into mere objects for the purpose of the law of armed conflict.

A given human enhancement technology may very well, however, constitute a weapon. This is the case when the technology is capable of causing either injury to or death of persons, or damage to or destruction of objects. Whether human enhancement technology possesses such offensive capabilities depends on the specific technology in question. On the one hand, a clear-cut instance where that question can be answered in the negative is biochemical enhancements of members of the State's own armed forces, since the enhancement itself does not cause injury to or death of enemy personnel nor damage to or destruction of objects. On the other hand, a possible instance where the answer is in the affirmative would be a prosthetic enhancement that integrates an offensive capability that is itself a weapon, such that the prosthetic is able to fire a munition or has the capability of a stun gun/taser.

Furthermore, human enhancement technologies may fall into the notion of means of warfare. An example would be a neural interface system through which weaponized drones are operated since the cybernetic system would form part of a weapons system employed for the purposes of attack.

Last, but not least, human enhancement technologies can be considered to be methods of warfare if and when their use constitutes an integral part of offensive activities at the strategic and tactical levels. In other words, if human enhancement technologies were to emerge as organizational features of, and techniques for, weapons and military units to deliver force against opposing armed forces and military objectives, they will trigger the review obligations under Article 36 of AP I and/or under domestic regulations.

What, then, are the main substantive principles of the law of armed conflict that would be utilized in such reviews and what idiosyncratic issues are raised when these principles are applied to human enhancement technologies?

---

[20]Program on humanitarian policy and conflict research (2009), r. 1(v).

[21]Sandoz et al. (1987), para. 1402; Boothby (2009), p. 4.

[22]See, e.g. Lin et al. (2013), pp. 31 ff.

2.2 *Superfluous Injury and Unnecessary Suffering*

The principle concerning superfluous injury and unnecessary suffering (SIrUS) has been recognized as one of the "cardinal principles" of international law by the International Court of Justice (ICJ).[23] First given form in the preamble to the 1868 Declaration of St Petersburg, the principle, which bans weapons, projectiles, materiel and methods of warfare that cause superfluous injury or unnecessary suffering, has evolved through its restatement in subsequent law of armed conflict treaties.[24] It is also now recognized as a rule of customary law applicable in both international and non-international armed conflicts.[25] However, the difference in language in consecutive legal instruments containing the prohibition[26] has made it difficult to determine whether the prohibition should be interpreted as prohibiting weapons "of a nature to cause" superfluous injury or unnecessary suffering (an objective standard) or whether it prohibits those weapons that are "calculated to cause" such injury. That is, the difference is between the necessary effects of the weapon and the intent of the user in designing and using it.[27]

It is this difference that may prove important not only when looking at the impact of the enhancement techniques themselves, but also on any countermeasures that will inevitably be developed in order to combat the military advantage gained through enhancement. For example, any weapon or technique designed to be used against enhanced soldiers which would cause greater suffering than is militarily necessary when employed against an unenhanced soldier would remain legal as its design and intended application is against enhanced soldiers. However, if one adopts an effects-based approach, the fact that the weapon may be used against unenhanced soldiers (perhaps without knowledge of their unenhanced status) would constitute a breach of the principle. The better view is that it is the intended design and normal effects of the weapon which is relevant to any assessment of the SIrUS principle.

---

[23] Legality of the Threat or Use of Nuclear Weapons, Advisory Opinion, 1996 I.C.J. 226, para. 238.

[24] For a description of the evolution of the different wording, see Hays Parks (2005), pp. 86 ff., The most recent versions include Art. 35(2) of Additional Protocol I which provides 'It is prohibited to employ weapons, projectiles and material and methods of warfare of a nature to cause superfluous injury or unnecessary suffering'. Identical wording is used in the preamble to the Convention on Prohibitions or Restrictions on the Use of Certain Conventional Weapons Which May be Deemed to be Excessively Injurious or to Have Indiscriminate Effects. Geneva, 10 October 1980.

[25] Henckaerts and Doswald-Beck (2005), Rule 70.

[26] Hays Parks (2005). See also Boothby (2009) pp. 55–61.

[27] An ill-fated project by the International Committee of the Red Cross (ICRC) in 1997 attempted to introduce an effects-based methodology which relied solely on medical evidence and, generally, the worst-case wounding effect of the weapon rather than the intended or normal effect. Hays Parks (2005). The ICRC project was heavily criticized for ignoring the balance to be struck with military necessity and was eventually withdrawn in 2001. Id.

A question also arises as to how the enhancements of soldiers should be viewed. For some enhancements, particularly those of a biochemical or genetic nature,[28] the question may be moot as there is no meaningful distinction between the soldier and the enhancement. However, for those enhancements which are cybernetic or prosthetic in nature, a distinction may be made between viewing the enhancement as a specific piece of technology separate from the soldier and as the soldier and technology combined into an integrated unit. To a certain extent this will be dependent on the nature of the enhancement technology in question. As noted above, the separation in the law of armed conflict between personnel and materiel means that enhanced soldiers themselves should not be viewed as weapons; however, that is different from saying that the enhancement should not be viewed as a means of warfare. Where the enhancement technology is intimately connected with soldiers' bodies and, in some cases, their brains, questions about the appropriate countermeasures that may be taken against those enhancements must be considered.

For example, if a brain-computer interface is viewed as a separate legitimate military objective, such that any physiological injury to the soldier would be viewed as incidental or collateral to the destruction or neutralization of the technology,[29] a different calculation may result when viewing the soldier and his or her enhancement as an integrated entity.

Clearly both the enhancement and the soldier may be considered legitimate targets. However, if it is permissible to kill the soldier or render him or her *hors de combat,* is it then automatically permissible to overload an implanted chip such that it may cause brain damage through an electrical overload of neural circuits? There are two issues at play in this situation. First, the SIrUS principle generally allows for more serious injuries resulting from anti-materiel weapons (for example, the use of depleted uranium shells in anti-tank weaponry or incendiary weapons against armored vehicles and fortifications) than it would for weapons typically designed for anti-personnel use.[30] This is because the military advantage resulting from destruction of the tank means the inevitable suffering of the tank's crew is not unnecessary. If one views the neural chip implanted in the brain of a cybernetically enhanced soldier as military materiel separate from the individual, the permissible level of injury and harm necessitated by the effects of a weapon designed to counter that chip may be higher than that of a weapon directed at unenhanced personnel. One might argue that any countermeasure used against implantable brain chips is designed to destroy the electronics in the implanted chip and any resulting brain damage to the soldier would be incidental. But, as Boothby notes in relation to the matter of the design, purpose and intent of blinding laser weapons, if it can be shown

[28]As noted above note 3, genetic enhancement is not generally addressed in this study however it is worth observing in passing that by definition, any genetic enhancement would not introduce any component external to the human body.

[29]Note that this does not refer to collateral damage in the sense of proportionality or precautions in attack, which refers only to civilians and civilian objects rather than military personnel (who, with limited exceptions, may be targeted at all times).

[30]See generally Greenwood (1998), pp. 185 and 196.

that a weapon will necessarily cause blindness when used in the circumstances for which it was procured, it may be difficult to show that it was not designed for that purpose.[31] Any determination would be highly fact-specific based on the technology employed, but a similar argument may be made for countermeasures which result in permanent brain injury or damage to neural circuitry.

A second point must also be considered. Abhorrence of particular types of injury has been the driver for many of the bans of specific weapons or uses of weapons as a method of war.[32] For example, blinding laser weapons were one of the few weapons that were prohibited before being fielded on the grounds that permanently blinding a soldier was superfluous and caused unnecessary suffering.[33] One can foresee that any countermeasure against implanted brain chips that would result in permanent brain damage might also fall within this category.

However, it should be borne in mind that the two elements of the prohibition on unnecessary suffering and superfluous injury are both comparative terms. Clearly a certain amount (and in some cases a great deal) of suffering and injury is permissible in armed conflict. What is prohibited is suffering or injury that is in excess of that necessary to realize a legitimate military goal (i.e., military necessity). Thus, if the only way to negate the threat posed by enhanced soldiers using implantable brain chips is to overload the chips, the resulting brain damage may in fact be necessary and the injury not superfluous.

Ironically, the prohibition against blinding laser weapons noted above is unlikely to apply to a soldier with enhanced vision. Certainly those with optical implants, or prostheses such as telescoping contact lenses, would not be covered by Article 1 of the Protocol on Blinding Laser Weapons which limits the prohibition to "unenhanced vision, that is to the naked eye or to the eye with *corrective* eyesight devices."[34] Article 3 specifically excludes from its ambit any blinding which takes place as an "incidental or collateral effect of the legitimate military employment of laser systems used against optical equipment." A question remains about those with vision that has been enhanced by other means. For example, Rain Liivoja raises the question of whether eyesight which has been augmented through surgical manipulation would fall within this definition (particularly if it would leave vision more susceptible to damage by laser).[35] Similar questions may be asked of eyesight which has been improved by other methods; for example, "biohackers" have been

[31] Boothby (2009), pp. 210 ff.

[32] See, e.g., Protocol on Prohibitions or Restrictions on the Use of Incendiary Weapons, Oct. 10, 1980, 1342 U.N.T.S. 171; Convention on the Prohibition of the Development, Production, Stockpiling and Use of Chemical Weapons and on their Destruction, Jan. 13, 1993, 1974 U.N.T.S. 45.

[33] Protocol on Blinding Laser Weapons, Oct. 13, 1995, 1380 U.N.T.S. 370.

[34] Emphasis added. A more difficult question may arise where the soldier's vision has been enhanced through genetic means, thus the soldier still relies on his or her naked eye. The answer would perhaps depend on the nature and degree of the enhancement.

[35] Liivoja (2015).

experimenting with increasing their night vision by using eye drops made from enzymes derived from deep-sea fish.[36]

2.3 *Distinction, Proportionality and Precaution Issues*

The principle of distinction requires that parties to an armed conflict distinguish between civilians and civilian objects on the one hand, and combatants and military objectives on the other; parties must only direct their military operations against the latter.[37] Along with the prohibition against unnecessary suffering (discussed above) it is considered one of the cardinal principles of international humanitarian law.[38] The principle of proportionality requires parties to refrain from, or cancel, attacks that are expected to cause incidental civilian injury or death, or damage to civilian objects (also known as collateral damage) which would be excessive in relation to the concrete and direct military advantage anticipated to be gained from the attack.[39] Compliance with these principles is impacted by both the environment and tempo of modern armed conflict, particularly in relation to information flows and current limits on the ability of humans to process such flows.

Today's armed forces deal with vastly more raw data and information than at any time in history.[40] Headsets, video feeds, instant messaging and radio transmissions all add to the cacophony of data that modern military forces are facing. This unprecedented amount of raw data helps by permitting more accurate targeting and in limiting collateral damage; however, these advantages must be balanced against the dangers of soldiers succumbing to the resulting "information overload," which can lead to tragic mistakes.[41] Work is ongoing in several research projects into increasing the brain's ability to multitask in order to be able to utilize the increased amount of data without becoming overwhelmed by it. These applied cognition methods range from attempts to train and "rewire" the functioning of the attention system to promote "mindfulness," to heads-up displays that monitor brain activity in

[36]The group dropped a chlorophyll analog (Chlorin e6) found in plants and some deep-sea fish into the eyeballs to give improved night vision. Alba (2015).

[37]See e.g. Art. 48 Additional Protocol I; Henckaerts and Doswald-Beck (2005), Rule 1, 7.

[38]Nuclear Weapons case, supra note 23, para. 78.

[39]The principle of proportionality runs as a thread through many of the substantive provisions of IHL, See e.g. Additional Protocol I, supra note 13, Arts. 51 (5) (b), 57 (2) (a) (iii), 57 (2) (b); Henckaerts and Doswald-Beck (2005), Rule 14.

[40]One estimate puts the increase in data flow across the U.S. armed forces at 1600% since the events of 9/11. Shanker and Richtel (2011).

[41]Id. When U.S. officials looked into an attack by American helicopters that left 23 Afghan civilians dead, they found that informational overload was an underlying cause. Id.

order to identify threats before they are recognized (or overlooked) by the conscious brain.[42]

If such methods are successful, the requirement to take all feasible precautions in attack, to ensure that any targets struck are legitimate military objectives, will necessarily be revised to take account of what is feasible when using the new technology. It should be noted that, as with past technological developments, most States consider that the law does not impose an obligation to acquire such technologies; however, if the State possesses the technologies it must field them when feasible.[43] In this sense, the use of human enhancement technologies may improve compliance with the requirements emanating from distinction and proportionality and their manifestation in the realm of precautions, but does not directly affect the content of the legal rules.

Another project currently under investigation by scientists is to inhibit or remove fear.[44] This is another instance in which the science may have interesting consequences for compliance with the principles of distinction, proportionality and precautions by allowing the soldier to make decisions without being motivated by fear. However, as noted, although such technologies may increase the likelihood of compliance with the existing laws, it will not impact the content of the legal principles themselves.

At the far (and perhaps speculative) end of the enhancement spectrum, exists the possibility of what Ingmar Persson and Julian Savulescu refer to as "moral enhancement."[45] Although fraught with ethical issues and of dubious practical use in the environment of an armed conflict, the theory of moral bio-enhancement suggests that by amplifying those biological factors which underlie a sense of justice and altruism, people will be compelled to act in a way that is morally (and, in the present context, legally) right.[46] The obvious difficulty would be ensuring that soldiers are trained to identify compliance with the law as the morally correct course of action, in contrast to increasing the sense of empathy (which is the course of action proposed by Persson and Savulescu in order to protect the environment) which may result in soldiers unfit for combat. It should be noted that in most

[42]See, e.g., Kaur (2012) (discussing the Defense Advanced Research Projects Agency's Cognitive Technology Threat Warning System program).

[43]Customary international law places a general requirement on States to avoid, and, in any event, minimize collateral damage. One of the ways in which this is done is through the choice of means and methods of attack. See generally Henckaerts and Doswald-Beck (2005), Rule 17 (discussing the requirement to take "all feasible precautions" to minimize incidental damage). See also, Dinniss (2012), pp. 213–215 (discussing new technologies and the choice of weapons as a feasible precaution).

[44]Moreno (2012), pp. 149–151.

[45]Persson and Savulescu (2012). While the concept of deliberately manipulating morality is speculative, it should be noted that the science behind it is not. A recent study found that common drugs used in the treatment of depression and Parkinson's disease sway moral decision making when administered to healthy people. Devlin (2015).

[46]In their original study, Persson and Savulescu (2012) argued for moral enhancement based on a need for protection of the natural environment, pp. 107 ff.

popular science fiction treatments of the topic, the opposite is usually postulated, i.e., the removal or minimization of empathy to create so-called “super-soldiers”; it invariably ends badly.[47] While a certain level of suppression of empathy is necessarily provided in combat training to produce effective combatants, permanent biochemical suppression beyond the individual soldier’s control risks producing troops that are unable to show the required levels of compassion and humanity for the wounded, sick or shipwrecked or for those who fall into their power, for example through surrender or as detainees.

2.4 *Principle of Protection (Detention and Treatment of the Wounded and Sick)*

The principle of protection requires that all persons who are not—or who are no longer—taking part in hostilities are treated humanely without any adverse distinction based on race, nationality, religious belief or political opinions, or any other distinction founded on similar criteria.[48] When considering enhancement technologies, one of the areas in which the principle raises particular issues is with regard to detention. These pertain primarily to the detainee who has been enhanced, but also in certain circumstances to the enhanced detainer. With regard to the latter, one can envisage situations where the use of particular enhancements by the detaining power on their own armed forces would reduce the possibility of abuses occurring through moral enhancement as discussed earlier or through methods such as optical recording (recording of the video stream of an implanted optical prosthetic could provide a view much like that of a headcam video feed) designed to increase rule compliance by detention personnel.[49] Although the use of such technology is likely to raise questions regarding the human rights of the personnel involved (particularly with regard to self-incrimination),[50] it does not implicate the laws of armed conflict per se. Broadly speaking, the law of armed conflict generally does not

[47]Lack of empathy is also associated with several psychological and personality disorders and is characteristic of psychopathy (or antisocial personality disorder). For examples in film and television see e.g. Robocop (Metro-Goldwyn-Mayer et al. 2014) or the character of Travis Verta in Continuum (Reunion Pictures & Shaw Media, 2012–2015), part of a super soldier program in which the character is implanted with a control chip that lowers empathy and compassion, whilst ramping up aggression, making him an unstable psychopath. For a general discussion of the popular culture trope of the ‘super soldier’ see http://tvtropes.org/pmwiki/pmwiki.php/Main/SuperSoldier.

[48]Henckaerts and Doswald-Beck (2005), Rule 88.

[49]It is well established that people comply with rules, and monitor and regulate their behavior when they believe that they are under surveillance. See, e.g. van der Linden (2011). Jeremy Bentham’s thought experiment of the panopticon and the subsequent discussion by Michel Foucault are classic treatments of the psychological phenomena. Foucault (1977), p. 201.

[50]See discussion infra Section 4.4.

concern itself with the protection of a State's own forces with regard to detention issues. This is subject to only limited exceptions in the area of the protection of the wounded, sick and shipwrecked in international armed conflicts (which extends to all those who are in need of care or assistance) and in non-international armed conflicts (where the protection of persons *hors de combat* extends to all members of armed forces).[51]

In applying the principle of protection to treatment of detainees, whether military or civilian, the law of armed conflict is directly implicated in a number of circumstances.

2.4.1 Coercion

Article 17 of the Third Geneva Convention (GC III) prohibits any form of coercion being inflicted on prisoners of war (POWs) in order to obtain information.[52] Thus, enhancing a prisoner's trust in his or her captors, by, for example, increasing their levels of oxytocin, a hormone tied to social bonding and sometimes referred to as the "cuddle hormone,"[53] would fall afoul of the sweeping and categorical prohibition of coercion. Coercion of protected persons, e.g., civilian detainees, is similarly prohibited under the Fourth Geneva Convention (GC IV).[54]

2.4.2 Experimental Treatments

Second, performing any invasive enhancement technique on detainees will be in breach of the laws regulating the treatment of both POWs and civilian detainees. Explicit treaty law found in GC III and GC IV and customary international law prohibit "[m]utilation, medical or scientific experiments or any other medical procedure not indicated by the state of health of the person concerned and not

[51]See generally Kleffner (2013), p. 285 (noting that the protection granted to the wounded, sick and shipwrecked extends to all those who are in need of medical assistance or care, provided they refrain from all acts of hostility).

[52]Convention (III) Relative to the Treatment of Prisoners of War art. 17, Aug. 12, 1949, 6 U.S.T. 3316, 75 U.N.T.S. 135 [hereinafter GC III] ("No physical or mental torture, nor any other form of coercion, may be inflicted on prisoners of war to secure from them information of any kind whatsoever. Prisoners of war who refuse to answer may not be threatened, insulted, or exposed to any unpleasant or disadvantageous treatment of any kind.").

[53]However, recent developments have also shown that oxytocin may have a dark side. Researchers have shown that negative or stressful social experiences, such as being bullied while certain structures in the brain have been activated by oxytocin, may make memories of that experience last long past the event itself and perhaps trigger fear and anxiety in the future. Bowman (2013).

[54]Convention (IV) Relative to the Protection of Civilian Persons in Time of War art. 31, Aug. 12, 1949, 6 U.S.T. 3516, 75 U.N.T.S. 287 [hereinafter GC IV]. It should be noted that this is not an absolute prohibition and exceptions exist where explicitly provided for in the Convention, for example in Articles 79–135, which regulate the treatment of internees.

consistent with generally accepted medical standards."[55] Article 11(2) of AP I further develops that prohibition, providing, *inter alia*, that it applies regardless of the consent of the person concerned. Article 5(2)(e) of Additional Protocol II contains broader language prohibiting "any medical procedure ... not consistent with generally accepted medical standards" with respect to persons detained in non-international conflict; however, no specific wording regarding consent is included. According to the International Committee of the Red Cross, an amalgamation of these conventional rules is also reflective of customary law applicable in both international and non-international armed conflicts.[56]

While the general rule is clear, specific scenarios may be envisaged where the use of enhancement technology raises interesting legal issues. One such issue turns on the experimental nature of the treatment. Whether or not a particular enhancement or treatment is deemed to be experimental has been the subject of review, particularly after the use of anthrax vaccinations by the U.S. armed forces in the 1991 Gulf War.[57] As a general rule, however, if the treatment has been approved by the treating State's standards authority, the procedure or treatment will not be viewed as experimental.[58]

2.4.3 Solving Medical Problems Through Enhancement

Another scenario raising interesting legal issues occurs when a detainee has a medical problem or injury that can best be resolved by a form of enhancement. One can imagine a situation in which wounded or sick enemy personnel require treatment and the standard treatment provided by the detaining power to its own forces would involve the use of an enhancement technique to block pain, promote rapid healing or enhance the immune system.

The United States' Defense Advanced Research Projects Agency (DARPA) currently has research programs in photobiomodulation to accelerate wound healing, vaccines that block pain within seconds and chemical cascades to stop bleeding

[55]Henckaerts and Doswald-Beck (2005), Rule 92. For specific treaty prohibitions, see Additional Protocol I, supra note 13, art. 11; GC III, supra note 52, art. 13; GC IV, supra note 54, art. 32; Protocol Additional to the Geneva Conventions of August 12, 1949, and Relating to the Protection of Victims of Non-international Armed Conflicts art. 5(2)(e), June 8, 1977, 1125 U.N.T.S. 609 [hereinafter Additional Protocol II].

[56]Henckaerts and Doswald-Beck (2005), Rule 92.

[57]Annas and Annas (2009), p. 283; Parasidis (2012), p. 1117.

[58]This function is performed for example by the Food and Drug Administration in the United States and Läkemedelsverket in Sweden. Rules relating to experimental treatments and trials are coordinated across the European Union and conducted under EU directives administered by the European Medicines Agency in accordance with standardized internationally recognized "good clinical practice" rules. See, e.g., Commission Directive 2005/28/EC of April 8, 2005, 2005 O.J. (L 91) 13 (laying down principles and detailed guidelines for good clinical practice as regards investigational medicinal products for human use) and incorporated directives.

within minutes.[59] In addition to these enhancements designed to address specific issues, DARPA has also been seeking nanoplatforms that can be introduced into the body for medical diagnostics, as well as being used to monitor and treat a broad range of diseases.[60] Although these programs are primarily focused on treatment rather than enhancement, the technologies described can last for a prolonged period of time (the effects of the pain vaccine last for 30 days) or can be used prophylactically (in the case of the diagnostic nanoplatforms), and thus fall within the enhancement paradigm. While these research programs are designed to ensure that U.S. personnel are diagnosed and treated effectively, their utility in maintaining the health of detainees confined in close quarters is immediately apparent.

As noted above, customary international law and Article 11 of AP I, protect any person who is in the power of an adverse party, or who is interned, detained or otherwise deprived of liberty as a result of an armed conflict. It is prohibited to subject such a person "to any medical procedure which is not indicated by the state of health of the person concerned and which is not consistent with generally accepted medical standards which would be applied under similar medical circumstances to persons who are nationals of the party conducting the procedure and who are in no way deprived of their liberty."[61] What constitutes generally accepted medical practice will be determined by medical ethics and the standards adopted by professional medical bodies, for example, the World Medical Association and the World Health Organization.[62]

In the case of the enhancement techniques being researched by DARPA, although the particular treatment would be indicated by the medical state of the detainee, intervention would not necessarily conform to generally accepted medical standards owing to its innovative status. However, if national authorities for the detaining State have approved the technology or technique for treatment of its own soldiers and/or citizens as a standard treatment and it conforms to the best medical interests of the patient, it may be argued that in those circumstances its use would be consistent with the law. In the commentary on the Copenhagen Guidelines on the Handling of Detainees in International Military Operations, it is noted that, although medical assistance should, wherever possible, be conducted with the consent of the wounded or sick detainee, medical actions to preserve the health of the detainee may be justified even where the detainee refuses to provide consent.[63] It should be noted that this is not the case with surgical interventions, where the person concerned may refuse an operation even if the surgeon considers it essential to the person's survival.[64]

---

[59] Garreau (2005), pp. 27–29.

[60] DARPA News (2012) and DARPA (2016).

[61] Additional Protocol I, supra note 13, art. 11(1).

[62] See, e.g., World Medical Association (1948, 1949, 1956). See also World Medical Association (1964) (addressing experimental treatments).

[63] The Copenhagen process: principles and guidelines (2012).

[64] Additional Protocol I, supra note 13, art. 11(5).

In the case of POWs, nanoplatforms designed for medical diagnosis may not only be permitted, but may also be preferred. Article 31, GC III requires that POWs undergo medical inspections at least once a month, specifically requiring that the most efficient methods available for the detection of contagious diseases, as well as for the supervision of the general health, nutrition and cleanliness of prisoners, be employed.[65] With technology costs inevitably decreasing over time, nanoplatforms may well become the most efficient way of monitoring the health of detainees.

2.4.4 Detaining Enhanced Personnel

A third scenario emerges when personnel who have previously been enhanced are detained. For example, if a detainee is reliant on periodic ingestion of a particular substance to maintain their enhancement, management of the withdrawal of the individual from that substance may be problematic. What legal consequences may flow from such a situation?

Under GC III, parties are required to repatriate seriously wounded or sick POWs to their own country.[66] This applies in particular to those whose mental or physical fitness seems to have been gravely diminished; those who are incurable; those who have recovered, but whose mental or physical fitness seems to have been gravely or permanently diminished; and those who require treatment and are unlikely to recover within a year.[67] Prisoners who enter detention while enhanced, and subsequently lose their enhanced capabilities, may appear to have a permanent diminishment in their physical or mental capabilities. For detaining forces with no knowledge of an individual soldier's baseline levels (i.e., the standard unenhanced level of whatever capability or function is being measured for that particular individual), forming an opinion about the difference between an enhanced soldier and their unenhanced state will necessarily be based on generalities. However, the purpose of the provisions is to facilitate the timely repatriation of gravely sick or wounded soldiers who can no longer return to active service. If the enhancement technology is such that without their drugs or other means soldiers will return to a normal baseline state of an unenhanced but otherwise fit-for-duty soldier, they would not qualify for repatriation. However should the enhancement be such that lack of maintenance would result in grave and permanent damage to the baseline status of the prisoner and the detaining power is unable to provide the appropriate

[65] GC III, supra note 52. Article 31 provides:

Medical inspections of prisoners of war shall be held at least once a month. They shall include the checking and the recording of the weight of each prisoner of war. Their purpose shall be, in particular, to supervise the general state of health, nutrition and cleanliness of prisoners and to detect contagious diseases, especially tuberculosis, malaria and venereal disease. For this purpose the most efficient methods available shall be employed, e.g. periodic mass miniature radiography for the early detection of tuberculosis.

[66] Id., art. 109.

[67] Id., art. 110.

treatment in its detention facilities, the prisoner must be transferred to a military or civilian medical facility where the appropriate treatment can be provided.[68]

3 Human Rights of Enhanced Humans

At the outset of this section, it is pertinent to make a point about the subject of *human* rights. Despite the enhanced status of members of the military or any other person who is the subject of an enhancement (whether pharmacological, cybernetic, genetic or otherwise), they remain, at their core, fundamentally human. The preamble of all the universal human rights instruments refer to "all members of the human family" and in their opening articles address the rights contained therein to "all human beings."[69] Like the universal instruments, regional human rights instruments also use inclusive language encompassing all persons or every human being,[70] or in the case of the European Convention on Human Rights (ECHR) secure rights to "everyone."[71] Lest the argument be made that a genetic change, technological implant or biological modification somehow remove the enhanced soldier from the human family, it should be recalled that the jurisprudence of human rights bodies, such as the European Court of Human Rights (ECtHR), have found breaches of protected rights of those whose brains are not yet fully formed (in cases involving an unborn child), those who have physiological differences, those who have genetic or chromosomal abnormalities, and those whose bodies have been changed by medical intervention.[72]

---

[68] Id., art. 30.

[69] See, e.g., Universal Declaration of Human Rights, G.A. Res. 217 (III) A, U.N. Doc. A/RES/3/217(III) (Dec. 10, 1948) [hereinafter UDHR]; International Covenant on Civil and Political Rights, Dec. 16, 1966, 999 U.N.T.S. 171 [hereinafter ICCPR]; International Covenant on Economic, Social and Cultural Rights, Dec. 16, 1966, 993 U.N.T.S. 3 [hereinafter ICESCR].

[70] See, e.g., American Convention on Human Rights, Nov. 22, 1969, 1144 U.N.T.S. 123 [hereinafter ACHR]; African Charter on Human and Peoples' Rights, June 27, 1981, 1520 U.N.T.S. 217 [hereinafter ACHPR].

[71] Convention for the Protection of Human Rights and Fundamental Freedoms art. 1, Nov. 4, 1950, 213 U.N.T.S. 222 [hereinafter ECHR].

[72] For example, the following European cases all feature complaints in which the subjects of the rights discussed are still regarded as rights holders. X v. United Kingdom, App. No. 8416/79, 19 Eur. Comm'n H.R. Dec. & Rep. 244 (1980) (observing that application of certain rights prenatally cannot be excluded); H.L. v. United Kingdom, 2004-IX Eur. Ct. H.R. 197 (complainant with severe autism and learning difficulties); Zarzycki v. Poland, App. No. 15351/03 (2013) (ECtHR), http://hudoc.echr.coe.int/eng?i=001-117210 (complainant with forearm prostheses); Tešic v. Serbia, App. Nos. 4678/07, 50591/12 (2014) (ECtHR), http://hudoc.echr.coe.int/eng?i=001-140771 (complainant with a pacemaker).

In determining the contours of human rights for members of the armed forces, the approach of the Inter-American human rights bodies[73] the European Union and the ECtHR has been to view them as citizens in uniform. The jurisprudence of the latter court is particularly rich in this respect, recognizing that human rights law applies in principle to members of the armed forces, but also acknowledging in a 1976 judgment that when interpreting and applying the ECHR "the Court must bear in mind the particular characteristics of military life and its effects on the situation of individual members of the armed forces."[74] The principle has been applied in subsequent cases, with the ECtHR observing that the extent of protection given to members of the armed forces must take into account the characteristics of military life, the nature of the activities they are required to perform and the risk arising from those activities.[75] This approach has led the ECtHR to provide the State with a considerable degree of latitude to restrict the human rights of the members of its armed forces during war or other public emergency threatening the life of the nation.[76] The Supreme Court of the United Kingdom, on reviewing the approach of the ECtHR to members of armed forces, observed:

> These comments, however brief, do seem to make it clear that it would not be compatible with the characteristics of military life to expect the same standard of protection as would be afforded by article 2(1) [of the ECHR] to civilians who had not undertaken the obligations and risks associated with life in the military. That is plainly so in the context of the exercise of military discipline over members of the armed forces when they are on active service. It is hard to see why servicemen and women should not, as a general rule, be given the same protection against the risk of death or injury by the provision of appropriate training and equipment as members of the police, fire and other emergency services. But it is different when the serviceman or woman moves from recruitment and training to operations on active service, whether at home or overseas. It is here that the national interest requires that the law should accord the widest measure of appreciation to commanders on the ground who have the responsibility of planning for and conducting operations there.[77]

From a practical standpoint, it should be noted that armed forces composed of volunteers, as opposed to conscript armies, face additional pressure to respect the human rights of their members in order to attract and retain the desired level and quality of recruits. Although an adult volunteer can be taken to have consented to certain aspects of military life and the resulting curtailment of particular rights by the very act of enlistment, the consent does not amount to a carte blanche waiver of all aspects of the human rights applicable to the individual.[78]

International human rights fall into four basic categories: non-derogable rights, rights which may be derogated from in war or public emergency threatening the

---

[73]See, e.g., J.S.C.H. & M.G.S., Case 12.689, Inter-Am. Comm'n on H.R., Report No. 80/15, OEA/Ser. L./V/II.156, doc. 33 (2015).

[74]Engel and Others v. Netherlands, 22 Eur. Ct. H.R. (ser. A) (1976).

[75]See, e.g., Grigoriades v. Greece, 1997-VII Eur. Ct. H.R.

[76]Rowe (2006), p. 59.

[77]Smith and Others v. Ministry of Defence [2013] UKSC 41, ¶ 71.

[78]Rowe (2006), pp. 9–13.

life of the nation, absolute rights and qualified rights. Qualified rights are rights which may be limited by law in certain specified circumstances. Both derogable rights and qualified rights may be significantly curtailed in respect of members of the armed forces. This is particularly the case in times of armed conflict as many of the limitation clauses contained in the relevant treaties specifically provide for limitations and restrictions for reasons of national security and/or public safety. Absolute rights (such as the prohibition against torture or cruel, inhuman or degrading treatment) remain in full force at all times, unaffected by membership in an armed force or the existence of an ongoing armed conflict or other situation affecting national security.

With these preliminary observations in mind, we will now address some of the human rights implications that flow from human enhancement technologies.

4 Human Rights Law Implications for Human Enhancement Technologies

It is clear that different enhancement technologies will impact the human rights of individual soldiers in different ways. This section briefly examines the impact of these technologies on some of the substantive rights set out in international human rights instruments.

4.1 The Right to Life

The right to life applies to soldiers both in peacetime and during armed conflict, although in the latter case the application is substantially curtailed by both jurisdictional and contextual matters. While members of the armed forces fall within the jurisdiction of their own State—whether serving at home or abroad—incidents involving third parties who are subjected to the actions of enhanced personnel deployed outside their own country will be governed by the rules on the extraterritorial application of human rights law, which in turn depend on whether an individual is subject to a State's jurisdiction.[79]

[79]Jurisdiction in the context of human rights law carries a special meaning, on which the jurisprudence of the ECtHR is still evolving. At present, for individuals to fall within the jurisdiction of a State, either the territory in which they are located, or the individuals themselves, must be under that State's effective control. Case law establishes that individuals held in detention by a State's armed forces are within the effective control of that State. Al-Skeini v. United Kingdom, 2011-IV Eur. Ct. H.R. 99. As a general proposition, the United States does not consider that human rights law applies extraterritorially: See e.g. Fourth periodic report of the United States to the UN Human Rights Committee at 142, U.N. Doc. CCPR/C/USA/4 (2012). That understanding has been reviewed in recent years to acknowledge the application of the Torture

The right to life is a non-derogable right, except in respect of deaths resulting from lawful acts of war.[80] However, as Rowe points out, "[t]he concept of lawful acts of war is concerned with the obligations a State owes to those (normally non-nationals) against whom it is engaged in an international armed conflict and not its own soldiers."[81] The right to life not only prohibits States from taking a life without justification, but also implies a duty on the part of the State to establish a framework of laws and procedures to protect life to the greatest extent reasonably possible under the circumstances.[82] It is in this obligation that the contextual restrictions on the right to life for members of the armed forces become pertinent. Soldiers are expected (and expect) to risk—and even sacrifice—their lives if necessary in the course of an armed conflict; however, this expectation does not entirely preclude liability on the part of the State with respect to the right to life of its own soldiers.

The UK Supreme Court had occasion to consider whether, and to what extent, the right to life in Article 2 of the ECHR imposes positive obligations on the government to prevent the deaths of its own soldiers.[83] The Court concluded that, while it must avoid imposing positive obligations on the State which are unrealistic or disproportionate in connection with the planning or conduct of military operations during armed conflict, it must give effect to those obligations where it would be reasonable to expect the individual to be afforded the protection of Article 2.[84] The Court then excluded from the ambit of the Article decisions taken at a high level of command that are closely linked to the exercise of political judgment and issues of policy, as well as decisions taken by those actively engaged in direct contact with the enemy. However, the Court was prepared to find there was room for claims to be brought in the middle ground, while acknowledging a wide margin of appreciation was to be granted to authorities and to those actively engaged in armed conflict.[85]

Convention to 'certain areas beyond' its sovereign territory, and more specifically to 'all places that the State party controls as a governmental authority,': U.N. Committee Against Torture (2014).

[80]See ICCPR, supra note 69, arts. 4, 6; ECHR, supra note 71, arts. 2, 15; ACHR, supra note 70, arts. 4, 27. The ECHR specifically exempts lawful acts of war under the derogations provision of article 15(2). In contrast, both the ICCPR and ACHR are expressed in terms of a prohibition of arbitrary deprivation of life; death arising from lawful acts of war are generally not considered arbitrary and therefore would not violate the right in the first place, rather than requiring a derogation.

[81]Rowe (2006), p. 137. It should be noted that this is not exclusively the case.

[82]The duty to take positive measures to protect the right to life derive both from specific statements in the relevant articles that the right should be protected by law and the general duty of States to ensure the rights recognized in the various instruments.

[83]Smith and Others supra note 77.

[84]Id., para. 76.

[85]The concept or doctrine of the 'margin of appreciation' describes the amount of latitude that the court will grant to states before disallowing measures put in place by the state that interfere with an individual's rights. The doctrine encapsulates the Court's recognition that the measures that a state may consider necessary to interfere with an individual's human rights may differ from state to state even within democratic societies. It also recognises that national governments are best placed to determine what is necessary for that state rather than an international court. Although primarily a function of the jurisprudence of the ECtHR, the rationale underlying the doctrine has

The Court concluded no hard and fast rules could be established; that each case would require an exercise of judgment based on its facts.[86]

The ECtHR itself has yet to consider a case based on the use of experimental technologies by the military, or indeed the extent to which the right to life in Article 2(1) offers any protection to armed forces members actively engaged in armed conflicts. However, the right to life must be balanced against the public interest of the State in conserving the fighting force. In the operational context, the State possesses a wide margin of appreciation in respect of its determination of the national interest. As noted above, the margin is significantly curtailed when personnel are engaged in training, as opposed to active warfighting.[87]

In relation to enhancement technologies, the right to life may be infringed where the use of a particular technology poses a risk to the individual soldier that results in his or her death. For example, fatal side effects of experimental enhancement technologies may involve an infringement of this right. Military experimentation, that is, the use of armed forces personnel as human test subjects, is replete with examples of horrific experiments conducted on often unknowing and/or objecting personnel.[88]

The right to life may also be impacted indirectly by enhanced soldiers who cause the wrongful death of others while under the influence of their enhancement, whether occurring during training or while engaged in military operations For example, in 2002, in what has become known as the Tarnack Farms friendly-fire incident, a U.S. Air Force F-16 pilot dropped a bomb on Canadian troops conducting a firing exercise, believing them to be Taliban fighters.[89] The pilots were returning to base following a 10-h night patrol when they reported what they believed to be surface-to-air fire. In the incident that followed, a combination of reckless behavior, bad communication and the fog of war led to one of the pilots dropping a 500 pound laser-guided bomb on the Canadian troops, killing four and wounding eight others. During the subsequent investigation and disciplinary proceeding, the pilots raised as part of their defense the fact that they had been told by superiors to use amphetamines during their mission and had taken dextroamphetamine prior to the incident. Although the defense was not accepted and the two pilots involved were held responsible for not following standard operating procedures and the rules of engagement, a number of medically qualified commentators have noted that the use of amphetamines for longer missions required of pilots by the Air Force is likely to

been applied by other international human rights bodies such as the IACHR and the HRC. See e.g. IACtHR, Proposed Amendments to the Naturalizations Provisions of the Constitution of Costa Rica, Advisory Opinion OC-4184 of January 19, 1984, ser. A, no. 4, para. 62; Hertzberg et al. v. Finland, U.N. GAOR Human Rights Comm., 37th Sess., Supp. No. 40 at 161, para. 10.3, U.N. Doc. A/37/40 (1982).

[86]Smith and Others, supra note 77, para. 76.

[87]See, e.g., id. paras. 67–76 (setting out the relevant case law of the Court).

[88]See generally Annas and Annas (2009); Parasidis (2012); Annas (1998), p. 245.

[89]For a summary of the incident, see Annas and Annas (2009), 47, pp. 293–297. As this only indirectly impacts the right to life, no further discussion will be included here.

have led to the pilots failure to wait for confirmation of the targets identity and to believe they needed to act in self-defense.[90] Not only are States responsible for the acts of their armed forces, but they also have an obligation where they know of a risk to the lives of those under their jurisdiction.[91]

It should be noted that, as with any other possible breach of the right to life, the right may also be infringed by a State's failure to investigate effectively.[92] Investigations must be independent, thorough and prompt, with a sufficient element of public scrutiny of the investigation or its results to ensure accountability.[93] These obligations are unchanged with respect to deaths that involve human enhancement technologies.

4.2 *Bodily Integrity, Torture, and Inhuman or Degrading Treatment*

The right of bodily integrity refers to the collection of rights (including the right to life) concerned with the inviolability of the human body and right of the person to determine what may be done to it. Referred to in slightly different terms under different treaty regimes, these rights incorporate security of the person,[94] the right to humane treatment,[95] the prohibition against slavery,[96] and the prohibition against torture and inhuman or degrading treatment,[97] as well as the prohibition against medical and scientific experimentation without freely given consent.[98] The prohibitions against slavery, and torture and inhuman or degrading treatment are absolute rights; States may not, under any circumstances, have recourse to such

[90]See, e.g., id., p. 296. Note, however, that others maintain that it should have improved the pilots' performance. Shanker and Duenwald (2003), http://www.nytimes.com/2003/01/19/national/19SPEE.html.

[91]See e.g. ECtHR, Osman v United Kingdom, EHRR 245 ¶¶ 115–22 (1998); IACtHR, González et al. (cotton field) v. Mexico, preliminary objection, merits, reparations and costs, Judgment of 16 November 2009, Ser. C, no. 205, ¶¶ 3–15 (concurring opinion of Judge Diego Garcia-Sayan).

[92]See Human Rights Committee, General Comment No. 6: (Right to Life), U.N. Doc. HRI/GEN/1/Rev. 7, at 129, ¶ 4 (2004).

[93]See generally Al-Skeini, supra note 79 (holding that deaths must be investigated effectively and impartially); in the IACtHR, Velásquez Rodríguez v Honduras, Merits, Judgment of 29 July 1988, Ser. C, no. 4, ¶¶ 172–7 (holding that states have the duty to investigate the right to life irrespective of the identity of the perpetrator).

[94]UDHR, supra note 69, art. 3.

[95]ACHR, supra note 70, art. 5.

[96]UDHR, supra note 69, art. 4; ICCPR, supra note 69, art. 8; ECHR, supra note 71, art. 4; ACHR, supra note 70, art. 6.

[97]ICCPR, supra note 69, art. 7; ECHR, supra note 71, art. 3; ACHR, supra note 70, art. 5(2).

[98]ICCPR, supra note 69, art. 7.

treatment or condone it.[99] States must, therefore, also take measures to protect its service personnel against such treatment.[100] These prohibitions, as well as the prohibition against medical and scientific experimentation without free consent, are also non-derogable rights under both the ICCPR and the ECHR.[101]

Enhancement of military personnel raises interesting questions with respect to the right of bodily integrity on two different levels. The first is the right of the soldier to consent to or refuse the enhancement and, second, what effects the enhancement may have on the permissible treatment of the soldier once enhanced.

The issue regarding the right to refuse an enhancement will largely depend on the nature and status of the technology involved. While approved medicines may become part of the standard pre-deployment medical regime in the same manner as conventional vaccines, more experimental technologies may raise legitimate concerns for the personnel involved. Indeed, many of the issues raised by the use of experimental technologies have been previously raised and debated with respect to the mandatory administration of anthrax vaccines which had not yet been approved by the States' drug approval agencies.[102] Whether a member of the armed forces has the right to refuse such vaccines will depend on the extent of his or her right to bodily integrity. The soldier who refuses the vaccine may be prevented from deploying with his or her unit as a potential risk to military efficiency, the lack of inoculation making him or her unfit for military service in a particular region. For example, in 2003 approximately 40 Australian servicemen and women who were deploying to the Middle East were ordered to return to Australia after refusing anthrax inoculations en route.[103] No disciplinary measures were taken in that case; however, in some States refusal has resulted in disciplinary action being taken against the soldier involved.[104]

Certainly where the enhancement remains experimental, the soldier will retain more rights to refuse the enhancement than if it has been approved for use by the appropriate State authority.[105] Regardless of the legal issues involved, medical ethics require informed consent of the subject for any experimental treatment.

[99]With respect to the obligation not to condone, see e.g., Art. 4–8 of the Convention against Torture requires states to make torture a crime under national laws and extradite or prosecute anyone present on their territory that is suspected of having committed an act of torture. States have a duty to investigate allegations of torture promptly, impartially and effectively (Arts. 12 and 13) and where substantial evidence is found, perpetrators of torture must be punished. Convention against Torture and Other Cruel, Inhuman or Degrading Treatment or Punishment Dec. 10, 1984, 1465 U.N.T.S. 85 [hereinafter CAT].

[100]Committee of Ministers (2010), https://wcd.coe.int/ViewDoc.jsp?id=1590149& Site=CM.

[101]ICCPR, supra note 69, art. 4(2); ECHR, supra note 71, art. 15.

[102]See, e.g., Rowe (2006), p. 47.

[103]Anthrax jab side-effects withheld (2004).

[104]See, e.g., Court-martial for refusing anthrax shot (2003).

[105]Such approval is granted in the United States by the Food and Drug Administration, the Medicines and Healthcare Products Regulatory Agency in the United Kingdom and the Läkemedelsverket (Medical Products Agency) in Sweden.

However, serious concerns have been raised by some authors who argue that true informed consent has been impossible to achieve in situations involving the military hierarchy,[106] citing examples where pressure (either real or perceived) from senior ranking officers for soldiers to comply with requests or, in some cases, direct orders to subject themselves to medical treatment, have removed any genuine ability to refuse.[107]

The right to bodily integrity is also often expressed as a function of the right to privacy. The jurisprudence of the ECtHR has provided leading judgments on the scope and contours of the right. The right to privacy requires that a State respect everyone's "private and family life, their home and correspondence."[108] The Court has noted that the term "private life" includes the physical and psychological integrity of a person, as well as "aspects of an individual's physical and social identity."[109] Courts in other jurisdictions have also viewed the right to refuse medical treatment as a function of the right to privacy.[110] Given that the right to privacy also concerns the physical integrity of a person, the right may be implicated where the armed forces require members to undergo enhancements as a requirement for deployment or to otherwise carry out their duties. Issues have arisen in the past where the State has required its soldiers deployed in an area in which it is suspected that the enemy may use chemical or biological weapons to take medically prescribed tablets or inoculations to protect them against the effects of such weapons.[111] The right to privacy has also been implicated in cases dealing with physical intrusions into the body, for example, in the anthrax medication cases discussed above.[112]

The right is not absolute either for civilians or members of the armed forces. For example, under the ECHR, States are permitted to restrict the right of privacy provided that the restriction is in accordance with the law, pursues a legitimate aim (for example, national security) and is necessary in a democratic society.[113]

---

[106]See Bird and Bird (2005), pp. 81–85; Annas (1998), p. 253, n. 42 (In the accompanying text the author points out that, although in principle the soldiers had the right to refuse a botulinum toxin vaccination, the right was not communicated to them in the field.).

[107]For example, the informed consent form used by the U.S. Air Force specifically states that although taking Dexedrine is voluntary, refusal to take the amphetamines may result in a pilot being grounded. Such an outcome would have a major impact on a pilot's career. See Naval Strike and Air Warfare Center (2000), p. 21.

[108]ECHR, supra note 71, art. 8. See also ICCPR, supra note 69, art. 17; ACHR, supra note 70, art. 11; Charter of Fundamental Rights of the European Union arts. 7–9, Dec. 18, 2000, 2000 O.J. (C 364).

[109]Pretty v. United Kingdom, 2002-III Eur. Ct. H.R. 154, ¶ 61.

[110]See, e.g., In re Quinlan, 70 N.J. 10, 40, 355 A.2d 647, 663 (1976) (basing a decision to terminate treatment, inter alia, on a constitutional right to privacy).

[111]Rowe (2006), p. 47.

[112]Id.

[113]ECHR, supra note 71, art. 8(2) ("There shall be no interference by a public authority with the exercise of this right except such as is in accordance with the law and is necessary in a democratic society in the interests of national security, public safety or the economic wellbeing of the country, for the prevention of disorder or crime, for the protection of health or morals, or for the protection

As to the permissible treatment of individual soldiers once enhanced, the subject of sleep deprivation provides an interesting example of the type of issue raised. Sleep deprivation has been denounced on multiple occasions by U.N. human rights bodies,[114] including the Committee against Torture, judicially recognized by the ECtHR as a form of inhuman and degrading treatment,[115] and prohibited as a method of interrogation by the Supreme Court of Israel,[116] among others.[117] In the case of the ECtHR, the Court recognized that sleep deprivation caused "intense physical and mental suffering,"[118] which surpassed the minimum level of severity to fall within the scope of inhuman treatment under Article 3 of the ECHR. However, the assessment of the minimum level of severity is relative and depends on all the circumstances of a case, such as the nature and context of the treatment, the manner and method of its execution, its duration, its physical and mental effects, and, in some cases, the gender, age and state of health of the victim.[119] Enhancement techniques which reduce the need for sleep or counteract the harmful effects of sleep deprivation in military personnel are seen as the holy grail of many defense-funded research projects. Should a technique be developed which allows personnel to require less sleep, or drugs developed which counteract the deleterious effects of sleep deprivation—lack of cognitive ability, mental agility etc.—it may call into question the accepted categorization of sleep deprivation as a prohibited technique constituting inhuman treatment or in some cases torture, at least in terms of the prolongation and levels of deprivation for soldiers who are so enhanced.

4.3 *Privacy, Thought and Expression*

Human enhancement technologies which mediate and/or manipulate information between the outside world and the brain (such as cybernetic implants or optical prosthetics) or which inhibit the brain's ability to make free choices (such as moral engineering) will also impact the human rights of those using them. In particular,

of the rights and freedoms of others."). Note that the jurisprudence of the court has also established that any interference with individual rights must also be proportionate to the legitimate aim sought to be realized. For an overview of the principle of proportionality as applied by human rights treaty bodies see Arai-Takahashi (2013), pp. 446–468.

[114]See, e.g., van Boven (2004) and U.N. Committee Against Torture (1997).

[115]Ireland v. United Kingdom, 25 Eur. Ct. H.R. (ser. A) at 58, para. 167 (1978).

[116]HCJ 5100/94 Public Committee Against Torture in Israel v. State of Israel 53(4) PD 817, paras. 31–32 (1999) (Isr.), reprinted in 38 International Legal Materials 1471 (stating that where sleep deprivation amounts to an end in itself for the purposes of breaking the suspect, rather than a side effect, it is prohibited).

[117]See, e.g., R (on the application of Binyam Mohamed) v. Secretary of State for Foreign and Commonwealth, [2010] EWCA Civ 65, United Kingdom: Court of Appeal (England and Wales).

[118]Ireland v. United Kingdom, supra note 115, para. 167.

[119]Id., para. 162; Soering v. United Kingdom, 161 Eur. Ct. H.R. (ser. A) at 32, para. 100 (1989).

the rights to privacy and freedom of thought, conscience, religion and expression find new challenges in these emerging technologies.

In the fields of cybernetics and prosthetics, privacy issues may also be raised by ocular implants or any other monitoring device which is connected to a system that has a data recording capability or provides for upload to a network or the cloud. While external devices such as helmet cameras or the dual-use Google Glass and its competitors can be removed by the user, implant technology cannot and would therefore have the capacity for 24/7 image capture or recording. Strict policies will need to be in place in developing the implant technology and interface and in its use to protect the privacy of both the user and those with whom they come in contact. Clearly, the recording and outside observation of private family moments will breach the soldier's right to a private family life.[120]

The Universal Declaration of Human Rights (UDHR), International Covenant on Civil and Political Rights (ICCPR) and some regional human rights treaties provide everyone (including members of the armed forces) the "freedom of thought, conscience and religion."[121] While it has thus far been exclusively used in cases dealing with freedom of religion and the associated right to express that religion, the ECtHR has noted "it is also a precious asset for atheists, agnostics, sceptics and the unconcerned."[122] Although specific limitations may be placed on the external manifestation of this right within the context of military life,[123] the protection in relation to the internal dimension of freedom of thought—the *forum internum*—is absolute and not subject to State interference. Indeed, Article 18(2) of the ICCPR specifically provides that "[n]o one shall be subject to coercion which would impair his freedom to have or to adopt a religion or belief of his choice." At first blush this right does not appear to impact enhancement technologies; however, the advent of brain-machine interfaces and the suggestion of moral enhancement raise the possibility of quite literally affecting a person's ability to think freely. Persson and Savulescu's controversial argument for moral enhancement of human beings goes

[120] Although privacy is a qualified, rather than absolute, right and therefore subject to the normal requirements of limitations on interferences with rights (lawful, necessary in a democratic society to achieve a legitimate aim and proportionate to that aim) it is difficult to see what legitimate aim might be served that would make routine outside observation of intimate family moments necessary or proportionate.

[121] UDHR, supra note 69, art. 18; ICCPR, supra note 69, art. 18; ECHR, supra note 71, art. 9. Other regional human rights instruments refer solely to freedom of conscience and religion. Although it should be noted that while the ACHR deals solely with conscience and religion in Article 12, freedom of thought is included with freedom of expression in the Article 13. ACHR, supra note 70. The ECHR provides that "[e]veryone has the right to freedom of thought, conscience and religion; this right includes freedom to change his religion or belief and freedom, either alone or in community with others and in public or private, to manifest his religion or belief, in worship, teaching, practice and observance." ECHR, supra note 71, art. 9.

[122] Kokkinakis v. Greece, 260 Eur. Ct. H.R. (ser. A) at 12, para. 31 (1993).

[123] As long as the restrictions comply with the requirements of, for example, Article 9(2), ECHR that they are properly prescribed by law, pursue a legitimate aim (e.g., national security) and are necessary in a democratic society.

directly to the very heart of the values protected by this right.[124] "At its most basic, … the right seeks to prevent state indoctrination of individuals by permitting the holding, development, and refinement and ultimately change of personal thought, conscience and religion."[125] If personnel are prevented from freely making moral judgments because of a drug administered to them by State authorities, their fundamental freedoms may well be violated.

While the current iteration of the problem sounds futuristic, the ECtHR has in fact had occasion to address the problem of thought control in the past. In the *Kokkinakis* judgment, the Court appeared to suggest that a breach of the right to freedom of thought, conscience and religion could occur in cases of severe spiritual coercion or brain-washing, although the Court suggested that it would also consider such acts to be in breach of Article 3 (prohibiting torture, inhuman or degrading treatment) of the ECHR.[126] The ECtHR has also recognized the particular vulnerability of military personnel to be influenced by senior officers by virtue of the military's hierarchical structure.[127]

Closely linked to the right of freedom of thought is the right to freedom of expression. For example, Article 10 of the ECHR provides: "Everyone has the right to freedom of expression. This right shall include freedom to hold opinions and to *receive and impart information and ideas without interference by public authority* and regardless of frontiers … ."[128] Other international and regional human rights instruments contain similar or broader wording.[129] Under the American Convention on Human Rights (ACHR), where freedom of thought and expression are contained in the same article, the Inter-American Court of Human Rights has recognized the closely linked "dual aspect" of the right,[130] and interpreted it more broadly as a right of substantive access to information, rather than mere prevention

[124]Persson and Savulescu (2012). As noted above, Persson and Savulescu make the claim in relation to environmental damage and climate change (although they also mention the threat of nuclear war), but there is the potential for application to the military in the moral enhancement of soldiers, particularly, although not exclusively, with those involved in detention after the scandals arising in the Iraq and Afghanistan conflicts. While abuse of detainees offers the best and most obvious example of immoral behavior, there are potentially many others which could affect the way soldiers operate, such as in targeting. However, enhancement of empathy could, in fact, make targeting decisions, particularly assessment of proportionality, harder on soldiers since it would be much more difficult for soldiers to accept and live with any collateral damage.

[125]Murdoch (2012), p. 18.

[126]Kokkinakis, supra note 122.

[127]Larissis and Others v. Greece, 1998-I Eur. Ct. H.R. 362. See also Martin et al. (2006).

[128]Emphasis added.

[129]See, e.g., UDHR, supra note 69, art. 19; ICCPR, supra note 69, art. 19; ACHR, supra note 70, art. 13; ACHPR, supra note 70, art. 9.

[130]Compulsory Membership in an Association Prescribed by Law for the Practice of Journalism (Arts. 13 and 29 American Convention on Human Rights), Advisory Opinion OC-5/85, Inter-American Court of Human Rights (ser. A) No. 5, ¶ 30 (Nov. 13, 1985).

of interference.[131] Needless to say, in respect of its application to members of the armed forces this right is often qualified.[132] Although none of the human rights instruments specifically mention the armed forces per se, most provide that the right may be restricted for reasons (among others) of national security.[133] The exception is most commonly used in the military context for such things as preventing publication of information (usually books or papers) by members of the armed forces that may compromise national security.[134] However, where soldiers are equipped with cybernetic implants (brain-machine interfaces) which mediate between an information source and the brain, the right to "receive and impart information without interference from a public authority" gains a new dimension.

There are many technologies which provide additional information to armed forces personnel, e.g., heads-up displays for fighter pilots and the Q-warrior augmented reality helmets from BAE Systems,[135] which are unlikely to impact this right. However, there are technologies in development which are intended to filter data in order to prevent information overload.[136] This may be particularly relevant where the implant or prosthetic removes visual information from view, or is designed to provide targeting information to the soldier. According to reports, software has been devised in Germany which allows for the deletion of visual information by smart glass or contact lens.[137] As one futurist was quoted as saying, "[s]o if you decide you don't like homeless people in your city, and you use this software and implant it in your contact lenses, then you won't see them at all."[138] One can imagine that States may be tempted to remove "extraneous" information from their already information-overloaded troops to enable them to concentrate on the task at hand.

While it is likely that the interference with this right will be governed by the national security exception (where available), law and policy makers should be aware of the impact of the technology on freedom of expression to ensure that a balance between the two is preserved. In particular, there should be a recognition that this type of technology will impact not only on the human rights of the personnel involved, but also the ability of those personnel to comply with such law of armed conflict principles as precautions in attack and proportionality.

---

[131] Rayes v. Chile, Merits, Reparations, and Costs, Judgment, Inter-American Court of Human Rights (ser. C) No. 151, paras. 61–103 (Sept. 19, 2006).

[132] Rowe (2006), pp. 55–58.

[133] Interestingly, this limitation does not occur in the ACHPR. Rowe (2006), p. 56. Additional limitations include that such restrictions must be prescribed by law and necessary in a democratic society.

[134] Rowe (2006), pp. 57–58.

[135] See McDuffee (2014).

[136] See, e.g., the DARPA-initiated augmented cognition program. Committee on Opportunities in Neuroscience for Future Army Applications (2009), p. 117, http://www.ncbi.nlm.nih.gov/books/NBK207983/.

[137] Morozov (2013) and Champion (2013).

[138] Khanna and Khanna (2016).

4.4 Fair Trial Rights

The right of everyone, including members of the armed forces, to a fair trial is provided by, *inter alia*, Articles 10 and 11(1) of the UDHR, Article 9 of the ICCPR, Articles 5 and 6 of the ECHR, Articles 8 and 25 of the ACHR, and Article 7 of the African Charter on Human and Peoples' Rights. These guarantees provide that in the determination of their civil rights and obligations, or of a criminal charge against them, everyone is entitled to a fair and public hearing within a reasonable time by an independent and impartial tribunal established by law. In addition to the specific rights enumerated in the text of the articles, additional guarantees have been developed through the jurisprudence of courts and tribunals.[139] Additional rights are also found elsewhere in the conventions: no punishment without law (Article 7, ECHR; Article 9, ACHR) and the right not to be tried or punished twice for the same offense (*ne bis in idem*) (Article 4, Protocol 7 to the ECHR[140]).

Military justice is a complex field in which a great variety of national systems operate.[141] States vary in the way they differentiate between military disciplinary matters and criminal offenses, in the type and composition of the courts dealing with military offenses, and in the putative independence of the staff investigating, prosecuting or reviewing cases vis-á-vis the chain of command. Each type of system raises different challenges for the human rights of armed forces members; however, the impact of human enhancement technologies on fair trial rights guaranteed under the various human rights treaties is the same regardless of the military or civilian status of the court. That said, there are certain technologies which may have an impact on fair trial rights that are more likely to be initially implemented by the military.

Articles 14(3)(g) of the ICCPR and 8(2)(g) of the ACHR contain an express right not to be compelled to testify against oneself or to confess guilt. Within the European system the right to silence and the privilege against self-incrimination are not found in the text of the ECHR, but have been developed through the case law of the ECtHR.[142] The Court has noted that "the ... right to silence and the right not to incriminate oneself are generally recognized international standards

---

[139] For example, under the European system these include the right to have access to a court, the right to remain silent and not incriminate oneself, the right to equality of arms and the right to an adversarial proceeding. See generally Human Rights Committee, General Comment 32, Article 14: Right to Equality before Courts and Tribunals and to a Fair Trial, U.N. Doc. CCPR/C/GC/32 (Aug. 23, 2007).

[140] Protocol 7 to the European Convention for the Protection of Human Rights and Fundamental Freedoms, Nov. 22, 1984, E.T.S. 117.

[141] For a description of different types of military justice systems and the human rights of the armed forces in the States participating in the Organization for Security and Co-operation in Europe, see Organization for Security and Co-operation in Europe, Office for Democratic Institutions and Human Rights (2008).

[142] See, e.g., K v. Austria, App. No. 16002/90, (ECtHR), http://hudoc.echr.coe.int/eng?i=001-57830 (noting the ECHR's right against self-incrimination).

which lie at the heart of a fair procedure under Article 6."[143] As always, it will depend on the nature of the technology involved; however, it is foreseeable that certain brain-machine interfaces and implants (such as ocular implants) may impact on this right for military personnel. For example, if an ocular implant is recording and transmitting everything that a soldier sees, a question arises over whether this will impact the privilege against self-incrimination in any subsequent proceeding regarding incidents portrayed in the footage. Generally speaking, the privilege against self-incrimination can operate to prevent the use of compelled information in subsequent criminal proceedings.[144] Objective evidence such as video footage (for example, from a helmet or body camera) does not fall within this right, as it is not communicative evidence and can be compelled to be provided by the user for a legitimate purpose. Arguably, however, one might draw a distinction between those implants that require some form of cognitive processing of the data by the individual and those which are merely a direct video feed and operate in much the same way as a helmet camera.

More traditional protections of this right, namely against the use of more directly coercive practices by the government to extract self-incriminatory evidence, will also adapt to the new technologies involved. Evidence obtained through the use of biochemical enhancers, such as oxytocin and other chemicals intended to make detainees more cooperative, will raise the same legal issues from a human rights perspective as the use of truth serums have in the past.[145] They are also likely to be open to the same criticisms, i.e., they make a person more susceptible to outside suggestion, thereby increasing the likelihood that the person would recall false memories and thus falsely incriminate themselves (or others). Use of biochemical enhancers would be specifically prohibited by the fundamental rights and freedoms contained in the Swedish Instrument of Government, which specifically provides protection against any "medical influence aimed at extorting or suppressing statements."[146] While the Swedish approach is generally considered to be reflective of customary law,[147] it is interesting to note that in March, 2013, a

[143]Heaney and McGuinness v. Ireland, 2000-XII Eur. Ct. H.R. 419, para. 40.

[144]Note that the exercise of this privilege differs across jurisdictions. In respect of the ICC see Rome Statute, Arts. 67(1)(g), 69(7); for an account of the English domestic legal tradition see Roberts and Zuckerman (2004), chpt 4.

[145]See, e.g. MacDonald (1955–1956), p. 259 (on the effectiveness of 'truth' serums generally); Human Rights Watch (2003); See generally, Jalloh v. Germany, 54810/00, European Court of Human Rights, 11 July 2006, paras. 103–123 (ruling that the use of a drug to compel evidence—in this case an emetic—violated the right to a fair trial).

[146]Regeringsformen [RF] [Constitution] 2, 5 (Swed.), reprinted in The Instrument of Government, Sveriges Rikssdag, https://www.riksdagen.se/en/How-the-Riksdag-works/Democracy/The-Constitution/The-Instrument-of-Government/ (then follow "The Instrument of Government" hyperlink) (last visited Feb. 10, 2016).

[147]See e.g. the broad wording of Art. 55(1)(b) Rome Statute which is considered reflective of customary international law and provides that a person "shall not be subjected to any form of coercion, duress or threat, to torture or to any other form of cruel, inhuman or degrading treatment or punishment;" (emphasis added).

judge in the United States approved the use of "truth serum" on James Holmes, the defendant in the Aurora, Colorado mass shooting incident.[148]

Other enhancements which affect memory recall will also create issues for fair trial rights. For example, research is ongoing into the use of beta-blockers and other pharmaceuticals to reduce the impact of traumatic memory formation.[149] Other proof-of-concept research is currently being carried out involving the manipulation of mouse memories to add emotional content.[150] Both projects have potential military application in the reduction of incidents of post-traumatic stress disorder in military personnel returning from combat tours. However, the fact that a soldier has undergone any form of memory manipulation will create doubt as to their reliability as a witness, either in their own defense or against another in a court action.

5 Human Rights Aspects of Reintegrating Enhanced Soldiers into Civilian Society

Once enhanced soldiers have completed their service and seek to return to civilian life, attention will have to be given to both the rights of the soldiers returning and to the wider societal impact of the presence of enhanced veterans in the population.

5.1 Removal of Enhancements

One of the major issues to be addressed will be what happens to the enhancements of enhanced soldiers once they leave the armed forces. Lin and his co-authors have raised the question of whether soldiers should be required to have any enhancements removed before integration back into civilian society so as not to give enhanced personnel an unfair advantage in the civilian workplace.[151] However, the legal situation regarding removal of enhancements is far from certain. For example, the ownership of medical devices (and even of the data received from them) differs across jurisdictions. From an ownership perspective, there may be a difference between implantable technology and prosthetics (albeit intimately connected ones). While some technologies may rightly be considered materiel of the armed forces and be required to be relinquished on exit from the armed forces—perhaps to

[148]Pilkington (2013). Truth serum was approved for the purposes of determining the insanity or otherwise of the accused in the event that he claimed an insanity defense.

[149]Giles (2005), pp. 448–449.

[150]Bilton (2013) (reporting that the scientists were able to add negative emotion to a neutral memory and make a mouse believe that it had been given an electric shock at a different location from that where the shock had actually occurred).

[151]Lin et al. (2013), p. 71. This issue is discussed in more detail in Sect. 5.2 infra.

be replaced by a more civilian-appropriate version, for example, in the case of a forearm prosthesis—other technologies may more properly be viewed as having become the property of the individual soldier.[152]

Careful research will be required to establish the physiological effects of the removal of any enhancements embedded within the body. For example, any cybernetic or brain-machine interface implanted in the brain may disrupt the normal functioning of neural pathways as the brain adapts to the new technology.[153] It is possible that subsequent removal of the implant may cause neurological damage if the brain is unable to reestablish its previous pathways or otherwise compensate for the loss of the technology. Thus, the removal of the enhancement could potentially violate the soldier's rights to bodily integrity and an adequate standard of healthcare,[154] or in severe cases amount to inhuman or degrading treatment.

In addition to any physiological harm, close attention must also be paid to the psychological effects of removing enhancements. Depending on the type of enhancement and the length of time the soldier has been using it, enhanced abilities may become part of the soldier's personal identity such that removal of them would be unduly traumatic. Any removal may impact the rights to bodily integrity and privacy (in respect of the soldier's psychological identity), and in some cases cruel, inhuman or degrading treatment. Thus care must be taken even when replacing a militarily-optimized enhancement for a more civilian-appropriate one to ensure similar functionality is provided. For example, a cutting-edge, mind-controlled forearm prosthetic, which has the ability to relate sensation back to the brain, should not be replaced with an inert mechanical claw.

5.2 Societal Rights and Discrimination

The concern about the impact of reintegrating military veterans into civilian society extends also to their treatment by mainstream society. If enhanced veterans were to present a significant threat to unenhanced workers in terms, for example, of lost job opportunities, the societal tendency to discriminate against perceived outsiders is likely to emerge. Of course, such a reaction will be based in part on the level of penetration of the particular technology from the military into the civilian society;

---

[152]To take a current example, pacemakers and implantable defibrillators become the property of the patient.

[153]Functional plasticity, that is the brain's (limited) ability to adapt and form new neural pathways, is well documented, particularly in instances of traumatic stress. Given that members of the armed forces are routinely deployed into situations which expose them to high levels of traumatic stress, this is of increased importance to military patients.

[154]UDHR, supra note 69, art. 25; ICESCR, supra note 69, art. 12. See also European Social Charter art. 13(1), Feb. 26, 1965, 529 U.N.T.S. 89 (ensuring a right to healthcare). Healthcare is understood to include the preservation of mental and physical health through medical services. See also Directorate General of Human Rights and Legal Affairs, Council of Europe (2010).

those enhancements which began as civilian technologies and were subsequently adopted by the military may be better accepted.

States have a duty to secure to their citizens (including members of the armed forces) their rights and freedoms without discrimination. For example, the UDHR provides for "equal protection against any discrimination in violation of this Declaration."[155] Article 26 of the ICCPR provides, "the law shall prohibit any discrimination and guarantee to all persons equal and effective protection against discrimination on any ground such as race, colour, sex, language, religion, political or other opinion, national or social origin, property, birth or other status." The ECHR contains almost identical wording with the addition of "association with a national minority" as a protected class.[156] The ACHR refers to "other social condition" rather than status.[157]

Discriminating against an enhanced veteran would not necessarily constitute prohibited grounds for discrimination under any of the applicable human rights instruments. However, while enhancement does not fall within any of the named categories, none of the lists of grounds for discrimination in the conventions are exhaustive.[158] In *Kjeldsen, Busk Madsen and Pedersen v. Denmark* the ECtHR suggested that "status" referred to "a personal characteristic by which persons or groups of persons are distinguishable from each other."[159] Although it may be dependent on the particular technology involved, enhancements may well come within that definition and fall into the "or other status" category set out in the conventions.

It should be noted that the principle of non-discrimination is not an absolute right. For example, it will not be violated if the distinction between individuals in analogous situations has an objective and reasonable justification, which must be assessed in relation to the aim and effects of the measures taken.[160] The aim must be legitimate and the measures proportionate to that aim. Thus, for example, a person denied employment because of an optical implant capable of image recording may not have a claim he or she has been discriminated against if the job involves working with sensitive or classified information. Preventing the risk that the implant may provide access to information to third parties is likely to be considered a legitimate objective and the decision to employ a person without such an enhancement is likely to be considered an appropriate and proportionate solution.

---

[155]UDHR, supra note 69, art. 7. See also Protocol No. 12 to the Convention for the Protection of Human Rights and Fundamental Freedoms art. 1, Nov. 4, 2000, E.T.S. No. 177 (creating a "free-standing" non-discrimination provision).

[156]ECHR, supra note 71, art. 14.

[157]ACHR, supra note 70, art. 1(1).

[158]See, e.g., Committee on Economic, Social and Cultural Rights (2009), paras. 15, 20–35 (discussing the need for a flexible approach to "other status" and further examples of prohibited grounds of discrimination including, inter alia, disability, nationality, age and health status).

[159]Kjeldsen, Busk Madsen and Pedersen v. Denmark, 23 Eur. Ct. H.R. (ser. A) at 24–25, para. 56 (1976).

[160]Human Rights of Members of the Armed Forces, supra note 141.

6 Accountability and Individual Criminal Responsibility

6.1 *State Responsibility*

A State considering the use of enhanced soldiers must also consider its responsibility for the acts of its organs under the doctrine of State responsibility.[161] The State remains responsible for acts carried out by State organs (including individuals) in their official capacity regardless of whether that State organ was acting *ultra vires* or contrary to instructions, or if the State knew of its actions.[162] The same rules apply to persons and entities that are empowered to exercise elements of governmental authority, but who are not State organs (for example, State contractors, such as prison guards, who are engaged to provide services).[163]

The rules of State responsibility deal with the responsibility of States for internationally wrongful acts attributable to them. It thus involves two elements: first, that the act is attributable to the State and, second, that the act constitutes a breach of an international obligation of the State.[164] Human enhancement technologies may play a role in determining whether or not this obligation has been breached. That is, an illegal act committed by an enhanced soldier might not be an internationally wrongful act because of the circumstances of his or her enhancement.

As a matter of general public international law, the law of State responsibility does not concern itself with subjective criteria such as the mental element, but is premised on the idea of objective liability.[165] It is the act or omission of the State which matters, regardless of any intent or fault. However, whether or not there has been a breach of a primary rule may depend on the intention or knowledge of the relevant State organs or agents.[166] For example, the crime of genocide requires the specified acts to be committed with the intent to destroy, in whole or in part, a particular group.[167] If the enhancement technology under consideration has destroyed the individuals' capacity to form the requisite intent,[168] it will not be

[161] While this article deals primarily with the impact of enhancement of the armed forces, this applies equally to members of the intelligence services and other state organs. To cite a fictional example, the main characters in the Bourne series of movies (played by Matt Damon & Jeremy Renner), featuring enhanced individuals, have both worked in different programs run by the intelligence services of the United States.

[162] Draft Articles on Responsibility of States for Internationally Wrongful Acts art. 7, Report of the International Law Commission, 53d Sess., Apr. 23–June 1, July 2–Aug. 10, 2001, U.N. Doc. A/56/10, GAOR 56th Sess., Supp. No. 10 (2001), reprinted in [2001] 2 Yearbook of the International Law Commission 32, U.N. Doc. A/CN.4/SER. A/2001/Add. 1 (Part 2).

[163] Id., arts. 5, 7.

[164] Id., art. 2.

[165] See generally Crawford (2013), pp. 60–61.

[166] Crawford (2002), pp. 81–82.

[167] Convention on the Prevention and Punishment of the Crime of Genocide art. 2, Dec. 9, 1948, 102 Stat. 3045, 78 U.N.T.S. 277.

[168] See Section 6.2 infra for a discussion of individual responsibility.

possible to establish that the breach of an international obligation by the State has occurred.[169]

Likewise with primary obligations containing a knowledge requirement; in the *Corfu Channel* case, Albania was held responsible to the United Kingdom for damage to two Royal Navy warships which struck mines laid in its territorial waters, because it knew or ought to have known of the mines' presence and failed to warn other States.[170] When dealing with the subject of human enhancement, and particularly with cybernetics, it must be asked what it means to know something.

Similar questions arise with a State that provides aid or assistance or direction and control in the commission of an internationally wrongful act by another State.[171] For the providing State to be held responsible, the aid, assistance, direction or control must be provided with knowledge of the unlawfulness of the subsequent act by the second State. For example, in the Bosnian *Genocide* case, the ICJ found that while there was little doubt that the atrocities committed in Srebrenica were committed with resources possessed as a result of the "general policy of aid and assistance" by Serbia and Montenegro, it was not established that Serbia was aware that the "perpetrators had the specific intent characterizing genocide."[172] The standard required by Article 16 (with respect to aiding and assisting) of the Draft Articles on State Responsibility of States is that the assisting State had specific knowledge, crucially, of an internationally wrongful act with a high degree of particularity,[173] and that it intended to facilitate the wrongful conduct through its aid or assistance. Therefore, if a State supplies enhancement technologies to another State for the purposes of enhancing their armed forces in such a manner that renders the soldiers incapable of determining the wrongfulness of their conduct, any subsequent breach of an international obligation by the receiving State would not engage the responsibility of the assisting State unless at the time the assistance was provided it had specific knowledge of the particular abuses or wrongful acts that would be committed.

6.2 *Individual Responsibility*

Questions of responsibility and autonomy may also be asked at the individual level. For example, if a soldier's fear or remorse for their past actions is removed or

[169]On the question of the need, or lack thereof, to attribute intent in the context of State responsibility for genocide, see Milanovic (2006), pp. 553–604 and 567–569; Nollkaemper (2003), p. 633.

[170]Corfu Channel (U.K. v. Alb.), Judgment, 1949 I.C.J. 4 (Apr. 9).

[171]See Draft Articles on Responsibility of States, supra note 162, arts. 16, 17.

[172]Application of the Convention on the Prevention and Punishment of the Crime of Genocide (Bosn. & Herz. v. Serb. and Montenegro), Judgment, 2007 I.C.J. 43, para. 422 (Feb. 26).

[173]Id., paras. 423–424. See also Dominicé (2010), p. 286.

diminished through pharmacological enhancement, what does that mean for the autonomy of the individual combatant? An essential and related question is how does that impact the individual's responsibility for any breaches of the law of armed conflict on the battlefield while so enhanced?

6.2.1 Free Will

At the extreme end of enhancement technology lays the risk that certain enhancements could interfere with the free will of the individual being enhanced. Such a determination would interfere with the individual's capacity to form the *mens rea* necessary to be held responsible for his or her actions under international criminal law. Article 30 of the Rome Statute of the International Criminal Court sets out the customary law standard for the mental element required when no specific rules regulate the *mens rea*,[174] indicating a person will only be held criminally liable for their acts where they are committed with intent and knowledge. The intent requirement can be divided into two parts. First, the person must intend to perform the specific act, i.e., they "mean to engage in the conduct" described in the crime.[175] Second, if the definition of the crime requires that a particular consequence results from the act (for example, the death of a person), the perpetrator must either "mean to cause that consequence" or at least be "aware that it will occur in the ordinary course of events."[176]

The use of human enhancement technologies may affect these requirements in a number of ways. For example, where prosthetics are controlled by cybernetic implants, such as an advanced military exoskeleton, the implant may be subject to interference either by the soldier's State or through a third party hacking the device. This would provide a defense similar to that of automatism, which occurs when the person involved proves it was an involuntary action over which he or she had no control.[177]

It is not just the control of prosthetics which may be compromised. Research conducted at the University of Washington on human-to-human interfaces enabled a researcher to control the hand movements of a colleague across campus.[178] As the experiment routed the control commands via the Internet, the controlled subject could have been geographically situated anywhere in the world, including a war zone. Control over the finger by the researcher involved in the study might equally be control over a soldier's finger on a weapon's triggering mechanism. This would

[174]Rome Statute of the International Criminal Court art. 30, July 17, 1998, 2187 U.N.T.S. 90. See also Werle (2005), pp. 104–106.

[175]Id.

[176]Id.

[177]See generally Werle (2005), p. 105 (noting that automatism prevents criminal responsibility for crimes under international law).

[178]Armstrong and Ma (2013).

negate the intention of the soldier not to perform the specific act, provided the will of the controller to move the finger can overrule the will of the soldier not to move it.

Likewise, moral enhancement, even if done for allegedly virtuous purposes, may well be seen as interfering with the free will of an individual to make their own choice by making the thought of acting in a contrary manner to what has been determined to be moral repugnant to them. The potential for abuse is obvious. Whether or not military personnel who have undergone such enhancement remain individually responsible for their actions will necessarily need to be determined on the basis of the enhancement technology and its capacity to interfere with their capability to appreciate the unlawfulness of their actions. In the case of specific instances of interference, forensic psychologists who deal with matters of free will in criminal cases not involving enhancements will need to be augmented by those with particular knowledge of the technologies concerned.

It is possible that enhancement technology may also impact on the knowledge requirement of individual responsibility. Where a crime requires the existence of certain circumstances, it is normally sufficient in establishing liability to prove that the perpetrator was aware that those circumstances existed.[179] For example, attacking a person in the knowledge that they are *hors de combat* requires merely an awareness of that person's inability to continue to participate in combat, rather than a correct legal assessment of their legal status. Where a crime requires knowledge of a descriptive material fact, sensory perception is all that is required. Naturally, enhancement techniques may either enhance or filter sensory perception, and will thus need to be taken into account when determining the individual's knowledge of the material facts, particularly where the relevant enhancement allows the user to interface directly with vast amounts of raw data.

Where the material fact in question is normative in nature, i.e., it requires a value judgment, it is sufficient if the perpetrator knows the fundamental factual circumstances and comprehends the significance or social relevance of the incriminating conduct.[180] It is possible that enhancement technologies such as moral enhancement may interfere with the ability of the individual to comprehend the social relevance of the facts; however, it seems likely that any such enhancement technique would be banned for that very reason before ever being fielded.

One final point to note is that many crimes contain a required specific mental element. For example, "*intentionally* directing attacks against the civilian population,"[181] "*willful* killing"[182] and "*willfully* causing great suffering"[183] under the Rome Statute. Likewise, its Elements of Crimes contain references to the mental

[179]Rome Statute, supra note 174, art. 30. See also Werle, supra note 153, at 105.

[180]For example, those elements of crime involving value judgements such as 'inhumane' or 'severe': Werle (2005), pp. 105–106; Triffterer (2008), p. 902.

[181]Rome Statute, supra note 174, art. 8(2)(b)(i) (emphasis added).

[182]Id., art. 8(2)(a)(i) (emphasis added).

[183]Id., art. 8(2)(a)(iii) (emphasis added).

element such as "*knew* or should have known"[184] and "was *aware* of,"[185] *inter alia*, which may be affected by the use of human enhancement technologies. Each of these mental elements will need to be assessed on a case-by-case basis in light of the impact of the particular technology on the individual's ability to meet those requirements.

6.2.2 Excused and Diminished Responsibility

The Rome Statute excludes a defendant from criminal responsibility where that person suffers from a "mental defect" or "was in a state of intoxication that destroys that person's capacity to appreciate the unlawfulness or nature of his or her conduct, or to control his or her conduct to conform to the requirements of the law."[186] The intoxication defense does not presuppose a specific substance (for example, both drink and drugs are covered); however, it is intended to apply only to external substances introduced into the body.[187] The use of pharmaceutical enhancements may fall within this category, as by analogy may other enhancements such as brain-machine interfaces where they have the same effect as that necessary for a mental defect or intoxication to provide a defense to a criminal charge. However, any enhancement which acts on the body's own functions, for example by adjusting body chemistry or genetics to create an ability within an individual to trigger increased levels of adrenaline or other neurochemicals at a later time, would not give rise to the defense of intoxication as the intoxicating agent would be a naturally formed (endogenic) substance.[188]

The defense of intoxication is not available in circumstances where the individual was voluntarily intoxicated when they either knew or disregarded the risk that they would likely engage in conduct constituting a crime.[189] There is a very real question to be considered on whether the taking of an enhancement drug which can mean success or failure of the mission—and in some cases literally the difference between life and death—can ever be considered truly voluntary in a hierarchical

[184]International Criminal Court, Elements of Crimes art. 8(2)(b)(vii), U.N. Doc. PCNICC/2000/1/Add. 2 (Nov. 2, 2000) (emphasis added).

[185]Id., art. 7(1)(e) (emphasis added). See Werle (2005), pp. 106–109.

[186]Rome Statute, supra note 174, art. 31(1)(b) (But excepting situations when "the person has become voluntarily intoxicated under such circumstances that the person knew, or disregarded the risk, that, as a result of the intoxication, he or she was likely to engage in conduct constituting a crime within the jurisdiction of the Court.").

[187]Triffterer (2008), pp. 876–877.

[188]Id. For an historical example, see the Viking berserker's legendary fighting frenzy which appeared to leave them impervious to pain and with heighted physical ability on the battlefield. There are different theories about its origins: ingestion of fly agaric mushrooms or the bog myrtle plant would speak to availably of the intoxication defense (albeit negated by its voluntary consumption), however a self-induced frenzy or hysteria (another of the theories) would not.

[189]Rome Statute, supra note 174, art. 31(1)(b).

military system. This echoes the arguments about true informed consent in the military setting.[190] There is certainly evidence of the pressure on U.S. troops to take drugs (both for performance enhancement in the case of Air Force pilots and prophylactically in the case of anthrax vaccines for ground troops in Iraq), indicating that the voluntary nature of such enhancements could be called into question.[191] Even if the defense is available, then clearly the question of whether the enhancement (whether pharmaceutical, implant or otherwise) affected the soldiers capacity or control to a sufficient degree such as to relieve them of criminal responsibility is a factual question that would have to be assessed on a case-by-case basis depending on the particular enhancement involved and the individual's reaction to it.

In either event, it is not sufficient that the defect or intoxication merely diminishes the person's capacity of appreciation or control, it must destroy the person's ability to realize that he or she is acting unlawfully or, if the person is aware of the unlawfulness of the conduct, lacks the ability to control it according to the requirements of the law.[192]

A more likely outcome in criminal trials where the use of enhancement technology by military personnel is implicated is the argument that the use of the enhancement diminishes, rather than negates, the responsibility of the personnel involved. The "special defense" of diminished responsibility is provided for in the Rules of Procedure and Evidence of the ICC,[193] and both the ICTY and the ICTR.[194] It provides for a plea of "substantially diminished mental capacity" as a mitigating circumstance in determining a sentence. However as the ICTY Appeals Chamber in Čelebići noted in relation to the insanity defence.

[190] Not only are the stakes much higher in terms of mission success, but the lives at stake are not only those of the individuals taking the drugs. Lin et al. (2013); Gross (2013), p. 92.

[191] The consent form for U.S. Air Force pilots using Dexedrine states that pilots may be grounded for 'safety reasons' in the event that they refuse to carry the pills, supra note 109. The two pilots in the Tarnak farms incident also stated that they felt pressure to take the pills, citing fear that they would be 'scrubbed from the mission' if they did not do so: Shanker and Duenwald (2003).

[192] Triffterer (2008), p. 547.

[193] Art. 78(1) of the ICC Statute and r. 145(2) of the ICC Rules of Procedure and Evidence: "In addition to the factors mentioned above, the Court shall take into account, as appropriate: (a) Mitigating circumstances such as: (i) The circumstances falling short of constituting grounds for exclusion of criminal responsibility, such as substantially diminished mental capacity or duress; . . . ".

[194] Rule 67(a)(ii)(b) of the Rules of Procedure and Evidence for both the ICTY and ICTR. "Within the time-limit prescribed by the Trial Chamber or by the pre-trial Judge appointed pursuant to Rule 65 ter: (i) the defence shall notify the Prosecutor of its intent to offer: (b) any special defence, including that of diminished or lack of mental responsibility".

7 Conclusion

Human enhancement techniques offer a tantalizing prospect for counteracting human frailty on the battlefield. For example, soldiers who need less sleep, can see clearly in the dark without bulky headwear or can run over rocky terrain for miles without getting fatigued, all offer measurable military advantages for the fighting force. However, the utilization of the more ambitious of these technologies also risks removing one of the real strengths of armed forces personnel—their humanity. The law of armed conflict allows soldiers to kill an enemy at one moment and obliges them to offer compassion and humane treatment in the next when that same enemy is not killed but is wounded or captured. Technologies which interfere with the ability to make that switch must be treated cautiously, taking into regard all the possible ramifications of their use. At the same time, there are not only risks associated with human enhancement techniques as far as compliance with the law of armed conflict is concerned, but also potential benefits. Techniques that would allow for the suppression or control of sentiments such as fear and revenge may also promote compliance. As with many other military technologies, the implications of human enhancement techniques for the law of armed conflict are not inherent in the development and fielding of the technology itself, but rather will depend on how they are designed and used. An equally serious legal issue to consider is that of the human rights implications of enhancement techniques for soldiers who are required or requested to undergo such enhancement, not only during their service in the armed forces, but also in their reintegration into society as they return to civilian life.

References

Alba A (2015) Group of biohackers test eye drops to give humans night vision. New York Daily News. http://nydn.us/1bZtIQa. Accessed 23 Nov 2016

Annas GJ (1998) Protecting soldiers from friendly fire: the consent requirement for using investigational drugs and vaccines in combat. Am J Law Med 24:245–260

Annas CL, Annas GJ (2009) Enhancing the fighting force: medical research on American soldiers. J Contemp Health Law Policy 25(2):283–308

Anthrax jab side-effects withheld (2004) BBC News, 21 February 2004. http://news.bbc.co.uk/2/hi/asia-pacific/3509037.stm. Accessed 23 Nov 2016

Arai-Takahashi Y (2013) Proportionality. In: Shelton D (ed) The Oxford handbook of international human rights law. Oxford University Press, Oxford, pp 446–468

Armstrong D, Ma M (2013) Researcher controls colleague's motions in 1st human brain-to-brain interface. UW Today, 27 August 2013. http://www.washington.edu/news/2013/08/27/researcher-controls-colleagues-motions-in-1st-human-brain-to-brain-interface/. Accessed 23 Nov 2016

Bienaimé P (2014) Mind-controlled drones are already a reality. Business Insider, 24 October 2014. http://www.businessinsider.com/drones-you-can-control-with-your-mind-2014-10?IR=T. Accessed 23 Nov 2016

Bilton N (2013) Computer-brain interfaces making big leaps. New York Times, 4 August 2013 at B4. http://bits.blogs.nytimes.com/2013/08/04/disruptions-rather-than-time-computers-might-become-panacea-to-hurt/?_r=0. Accessed 23 Nov 2016
Bird J, Bird G (2005) Human rights and the military: the "chemical soldier". Alternative Law J 30(2):81–85
Black T (2013) Iron Man meets HULC as Lockheed enters exoskeleton race. Bloomberg Technology, 19 March 2013. https://www.bloomberg.com/news/articles/2013-03-19/iron-man-meets-hulc-as-lockheed-enters-exoskeletons-race. Accessed 23 Nov 2016
Boothby WH (2009) Weapons and the law of armed conflict. Oxford University Press, Oxford
Bowman L (2013) Oxytocin studies show Hormone's Dark Side. ABC 10 News, 28 July 2013. http://www.10news.com/lifestyle/health/medical-oxytocin-studies-show-hormones-dark-side-07282013
Champion E (2013) Thirty-five arguments against Google Glass. Reluctant Habits, 14 March 2013. http://www.edrants.com/thirty-five-arguments-against-google-glass/. Accessed 30 Nov 2016
Choi C (2015) New army exoskeleton makes soldiers' gunshots more accurate. Popular Science, 2 June 2015. http://www.popsci.com/army-has-exoskeleton-makes-soldiers-better-shots. Accessed 23 Nov 2016
Committee of Ministers (2010) Human rights of members of the armed forces. Recommendation CM/Rec (2010) 4, 24 February 2010. https://search.coe.int/cm/Pages/result_details.aspx?ObjectID=09000016805cf8ef. Accessed 23 Nov 2016
Committee on Economic, Social and Cultural Rights (2009) General comment No. 20: non-discrimination in economic, social and cultural rights (art. 2, para. 2, of the International Covenant on Economic, Social and Cultural Rights). U.N. Doc. E/C.12/GC/20
Committee on Opportunities in Neuroscience for Future Army Applications (2009) Opportunities in neuroscience for future army applications. https://doi.org/10.17226/12500
The Copenhagen process: principles and guidelines (2012) The Copenhagen process on the handling of detainees in international military operations. http://um.dk/en/~/media/UM/English-site/Documents/Politics-and-diplomacy/Copenhangen%20Process%20Principles%20and%20Guidelines.pdf. Accessed 1 Dec 2016
Court-martial for refusing anthrax shot (2003) New York Times, 5 April 2003. http://www.nytimes.com/2003/04/05/nyregion/court-martial-for-refusing-anthrax-shot.html. Accessed 23 Nov 2016
Crawford J (2002) The International Law Commission's articles on state responsibility: introduction, text, and commentaries. Cambridge University Press, Cambridge
Crawford J (2013) State responsibility: the general part. Cambridge University Press, Cambridge
Daniels N (2000) Normal functioning and the treatment-enhancement distinction. Camb Q Healthc Ethics 9:309–322. https://doi.org/10.1007/s10943-014-9882-7
DARPA (2016) In Vivo Nanoplatforms (IVN). http://www.darpa.mil/program/in-vivo-nanoplatforms. Accessed 23 Nov 2016
DARPA News (2012) DARPA effort targets illness faster, safer and more effectively. S&T News Bulletin, 15 June 2012, 2(24). http://www.acq.osd.mil/chieftechnologist/publications/docs/ST_NewsBulletin-V2-I24(15JUN2012).pdf. Accessed 23 Nov 2016
Devlin H (2015) Parkinson's and depression drugs can alter moral judgment, study shows. The Guardian, 2 July 2015. https://www.theguardian.com/science/2015/jul/02/parkinsons-and-depression-drugs-can-alter-moral-judgement-study-shows. Accessed 23 Nov 2016
Dinniss HH (2012) Cyber warfare and the laws of war. Cambridge University Press, Cambridge
Directorate General of Human Rights and Legal Affairs, Council of Europe (2010) Human rights of members of the armed forces. CM/Rec (2010) 4. http://www.coe.int/t/dghl/standardsetting/hrpolicy/publications/cmrec_2010_4en.pdf. Accessed 23 Nov 2016
Dominicé C (2010) Attribution of conduct to multiple states and the implication of a state in the act of another state. In: Crawford J, Pellet A, Olleson S (eds) The law of international responsibility, 1st edn. Oxford University Press, Oxford, pp 281–289
Drummond K (2012) Prosthetics breakthrough might fuse nerves with fake limbs. Wired, 27 February 2012. https://www.wired.com/2012/02/nerve-prosthetics/. Accessed 23 Nov 2016

Foucault M (1977) Discipline and punish (trans: Sheridan A). Vintage Books, New York

Garreau J (2005) Radical evolution: the promise and peril of enhancing our minds, our bodies—and what it means to be human. Broadway Books, New York

Giles J (2005) Beta-blockers tackle memories of horror. Nature 436(7050):448–449

Greenwood CJ (1998) The law of weaponry at the start of the new millennium. In: Schmitt MN, Green LC (eds) The law of armed conflict: into the next millennium, U.S. naval war college international law studies, vol 71. Naval War College, Newport, pp 185–231

Gross ML (2013) Military medical ethics: a review of the literature and a call to arms. Camb Q Healthc Ethics 22(1):92–109

Hays Parks W (2005) Conventional weapons and weapons reviews. Yearb Int Humanit Law 8:55–142. https://doi.org/10.1017/S1389135905000553

Heinlein RA (1959) Starship troopers. Putnam, New York

Henckaerts J-M, Doswald-Beck L (eds) (2005) Customary international humanitarian law. Cambridge University Press, Cambridge

Human Rights Watch (2003) The legal prohibition against torture. https://www.hrw.org/news/2003/03/11/legal-prohibition-against-torture#serums. Accessed 30 Nov 2016

Huxley A (1932) Brave new world. Chatto & Windus, London

Johnson JK (2012) Super-history: comic book superheroes and American Society. McFarland, Jefferson

Kaur K (2012) New sensor system improves target detection. AZO Robotics, 21 September 2012. http://www.azorobotics.com/News.aspx?newsID=3237. Accessed 23 Nov 2016

Khanna P, Khanna A (2016) The pleasure and danger of augmented reality. Big Think. http://bigthink.com/hybrid-reality/the-pleasure-and-danger-of-augmented-reality. Accessed 30 Nov 2016

Kleffner JK (2013) Friend or foe? On the protective reach of the law of armed conflict. In: Matthee M, Toebes B, Brus M (eds) Armed conflict and international law: in search of the human face. T.M.C. Asser Press, The Hague, pp 285–302

Liivoja R (2015) Bioenhanced soldiers and the prohibition of unnecessary suffering. Remarks at the Swedish Defence University, Stockholm, 9 Sept 2015

Lin P (2010) Ethical blowback from emerging technologies. J Mil Ethic 9(4):313–331. https://doi.org/10.1080/15027570.2010.536401

Lin P, Mehlman MJ, Abney K (2013) Enhanced warfighters: risk, ethics, and policy. Case Legal Studies Research Paper No. 2013-2. http://ethics.calpoly.edu/greenwall_report.pdf. Accessed 30 Nov 2016

MacDonald JM (1955) Truth serum. J Crim Law Criminol Police Sci 46:259–263

Martin FF, Schnably SJ, Wilson RJ, Simon JS, Tushnet MV et al (2006) International human rights and humanitarian law: treaties, cases and analysis. Cambridge University Press, Cambridge, p 747

McClelland J (2003) The review of weapons in accordance with Article 36 of Additional Protocol I. IRRC 85(850):97–415

McDuffee A (2014) At last, a Google Glass for the Battlefield. Wired, 24 February 2014. http://www.wired.com/2014/02/battlefield-glass/. Accessed 1 Dec 2016

Metro-Goldwyn-Mayer (MGM), Columbia Pictures, Strike Entertainment, Dynamic Effects Canada (Production Companies), Padilha J (Director) (2014) RoboCop. MGM, United States

Milanovic M (2006) State responsibility for genocide. Eur J Int Law 17(3):553–604. https://doi.org/10.1093/ejil/chl019

Moreno JD (2012) Mind wars: brain science and the military in the 21st century. Bellevue Literary Press, New York

Morozov E (2013) The perils of perfection. New York Times, 3 Mar 2013, at SR1. http://www.nytimes.com/2013/03/03/opinion/sunday/the-perils-of-perfection.html?_r=0. Accessed 1 Dec 2016

Murdoch J (2012) Protecting the right to freedom of thought, conscience and religion under the European Convention on Human Rights. Counsel of Europe Handbooks. http://www.coe.int/t/dgi/hr-natimplement/Source/documentation/hb09_rightfreedom_en.pdf

Naval Strike and Air Warfare Center (2000) Performance maintenance during continuous flight operations: a guide for flight surgeons (NAVMED P-6410). http://www.med.navy.mil/directives/Pub/6410.pdf

Nollkaemper A (2003) Concurrence between individual responsibility and state responsibility in international law. Int Comp Law Q 52(3):615–640

Office of the General Counsel, U.S. Department of Defense (2015) Law of war manual

Organization for Security and Co-operation in Europe, Office for Democratic Institutions and Human Rights (2008) Handbook on human rights and fundamental freedoms of armed forces personnel. Agencja Karo, Warsaw. http://www.osce.org/odihr/31393?download=true

Parasidis E (2012) Human enhancement and experimental research in the military. Conn Law Rev 44(4):1117–1132

Persson I, Savulescu J (2012) Unfit for the future: the need for moral enhancement. Oxford University Press, Oxford

Pilkington E (2013) Judge approves use of "truth serum" on accused aurora shooter James Holmes. The Guardian, 12 March 2013. https://www.theguardian.com/world/2013/mar/12/judge-approves-truth-serum-james-holmes. Accessed 1 Dec 2016

Program on humanitarian policy and conflict research (2009) Manual on international law applicable to air and missile warfare. Bern, 15 May 2009. http://ihlresearch.org/amw/HPCR%20Manual.pdf

Program on humanitarian policy and conflict research (2010) Commentary on the manual on international law applicable to air and missile warfare. http://ihlresearch.org/amw/Commentary%20on%20the%20HPCR%20Manual.pdf

Roberts P, Zuckerman A (2004) Criminal evidence. Oxford University Press, Oxford

Rowe PJ (2006) The impact of human rights law on armed forces. Cambridge University Press, Cambridge

Saletan W (2013) The war on sleep. Slate, 29 May 2013. http://www.slate.com/articles/health_and_science/superman/2013/05/sleep_deprivation_in_the_military_modafinil_and_the_arms_race_for_soldiers.html. Accessed 1 Dec 2016

Sandoz Y, Swinarski C, Zimmermann B (eds) (1987) Commentary on the additional protocols of 8 June 1977 to the Geneva Conventions of 12 August 1949. Martinus Nijhoff, Leiden

Shanker T, Duenwald M (2003) Bombing error in Afghanistan puts a spotlight on pilots' pills. New York Times, 19 January 2003. http://www.nytimes.com/2003/01/19/us/threats-and-responses-military-bombing-error-puts-a-spotlight-on-pilots-pills.html. Accessed 1 Dec 2016

Shanker T, Richtel M (2011) In new military, data overload can be deadly. New York Times, 16 January 2011. http://www.nytimes.com/2011/01/17/technology/17brain.html?pagewanted=all&_r=0. Accessed 1 Dec 2016

Sherwell P (2015) Blind man describes joy at seeing wife for first time in decade thanks to 'bionic eye'. The Telegraph, 24 February 2015. http://www.telegraph.co.uk/news/worldnews/northamerica/usa/11433149/Blind-man-describes-joy-at-seeing-wife-for-first-time-in-decade-thanks-to-bionic-eye.html. Accessed 1 Dec 2016

Talbot D (2013) An artificial hand with real feelings. MIT Technology Review, 5 December 2013. https://www.technologyreview.com/s/522086/an-artificial-hand-with-real-feelings/. Accessed 1 Dec 2016

Tennison MN, Moreno JD (2012) Neuroscience, ethics, and national security: the state of the art. PLoS Biol 10(3). https://doi.org/10.1371/journal.pbio.1001289

Thomson H (2015) Narcolepsy medication modafinil is world's first safe "smart drug". The Guardian, 19 August 2015. https://www.theguardian.com/science/2015/aug/20/narcolepsy-medication-modafinil-worlds-first-safe-smart-drug. Accessed 1 Dec 2016

Triffterer O (ed) (2008) Commentary on the Rome statute of the international criminal court, 2nd edn. C.H. Beck, Munich

U.N. Committee Against Torture (1997) Report of the Committee against Torture. U.N. Doc. A/52/44 (Supp)

U.N. Committee Against Torture (2014) Concluding observations on the combined third to fifth periodic reports of the United States of America. U.N. Doc. CAT/C/USA/CO/3-5

U.N. Human Rights Committee (2012) Concluding observations on the fourth periodic report of the United States. U.N. Doc. CCPR/C/USA/4

U.S. Deputy Secretary of Defense (2003) The defense acquisition system, DoDD 5000.01. http://www.dtic.mil/whs/directives/corres/pdf/500001p.pdf. Accessed 1 Dec 2016

Ungerleider N (2012) DARPA's cybernetic binoculars tap soldiers' brains to spot threats. Fast Company, 21 September 2012. http://www.fastcompany.com/3001501/darpas-cybernetic-binoculars-tap-soldiers-brains-spot-threats. Accessed 1 Dec 2016

Universal Pictures (Production Company), Gilroy T (Director) (2012) The Bourne Legacy [motion picture]. Universal Pictures, United States

van Boven T (2004) Special rapporteur on torture and other cruel, inhuman or de-grading treatment or punishment. In: Sixth report pursuant to General Assembly Resolution 58/164 and Human Rights Commission Resolution 2004/41, 1 September 2004. U.N. Doc. A/59/324

van der Linden S (2011) How the illusion of being observed can make you a better person. Scientific American, 3 May 2011. http://www.scientificamerican.com/article/how-the-illusion-of-being-observed-can-make-you-better-person/. Accessed 30 Nov 2016

Werle G (2005) Principles of international criminal law. T.M.C. Asser Press, The Hague

World Medical Association (1948) Declaration of Geneva. http://www.wma.net/en/30publications/10policies/g1/. Accessed 1 Dec 2016

World Medical Association (1949) International code of medical ethics. http://www.wma.net/en/30publications/10policies/c8/. Accessed 1 Dec 2016

World Medical Association (1956) Regulations in times of armed conflict and other situations of violence. http://www.wma.net/en/30publications/10policies/a20/. Accessed 1 Dec 2016

World Medical Association (1964) Declaration of Helsinki—ethical principles for medical research involving human subjects. http://www.wma.net/en/30publications/10policies/b3/. Accessed 1 Dec 2016

Yaron M (2016) Statement on Lethal Autonomous Weapons Review Systems (LAWS). Weapon Legal Review, address before the Group of Experts Meeting on Lethal Autonomous Weapons Systems, 13 April 2016. http://www.unog.ch/80256EDD006B8954/(httpAssets)/A02C15B2E5B49AA1C1257F9B0029C454/$file/2016_LAWS_MX_GeneralDebate_Statements_Israel.pdf. Accessed 1 Dec 2016

Meaningful Human Control – and the Politics of International Law

Thilo Marauhn

Abstract In the Geneva discussions on lethal autonomous weapons systems the concept of "meaningful human control" plays an important role. While the concept as such has served to focus on particular politico-ethical questions and to place the issue of lethal autonomous weapons systems prominently on the agenda of the High Contracting Parties to the Convention on Certain Conventional Weapons (CCW), its legal value is questionable. This chapter highlights the importance of terminology—in law and in politics. It illustrates that terminology can serve different purposes in political and legal contexts when negotiating questions of international humanitarian law (and arms control). The chapter illustrates that the concept of "meaningful human control" does not add anything to existing standards of the law of armed conflict, perhaps even risks blurring them, but should in the first place be used for political purposes. Addressing time and place of decision-making, cognition and volition, as well as the relevance of value-based decision-making, the chapter concludes that the concept may, nevertheless, facilitate the implementation of existing law.

1 Introduction

Terminology matters—in law and in politics. This is particularly true for the international level, where terminology is an important element of the politics of law. The concept of "meaningful human control" is a pertinent example in the context of international security.

In order to preserve and enhance the security of their citizens, states can no longer confine themselves to address threats arising from other nation states. They are

T. Marauhn, Professor for Public Law and International Law, Justus Liebig University Gießen, Germany. I am grateful to Barry de Vries for valuable comments and assistance in the drafting of this chapter.

T. Marauhn (✉)
Public Law and International Law, Justus Liebig University Gießen, Gießen, Germany
e-mail: thilo.marauhn@recht.uni-giessen.de

W. Heintschel von Heinegg et al. (eds.), *Dehumanization of Warfare*,
DOI 10.1007/978-3-319-67266-3_11

expected to ensure protection also against non-state actors whatever their specific characteristics and legal nature may be. This has nurtured, among others, the incorporation of remote piloting, automated attack technology and cyber techniques into the array of instruments applied by states in order to address perceived threats more effectively and to better protect their own forces than in direct combat operations.[1]

In turn, this has stimulated a debate about the (political) desirability, the legitimacy, and the legality of automated and autonomous weapons and weapon systems. Apart from many others, the question has arisen to what extent "de-humanizing warfare" is compatible with international humanitarian law *de lege lata* and whether there is a need to prohibit or restrict the use of pertinent weapons and weapon systems, including by means of preventive arms control.[2]

Negotiations in respect of the latter are being conducted in the framework of the Convention on Certain Conventional Weapons (CCW).[3] Without looking at the prospects of these negotiations and the viability of a possibly emerging document on so-called lethal autonomous weapons (unfortunately often abbreviated as LAWS, with all the ambiguity and ambivalence attached), the following remarks focus on the framing of the debate under the heading of "meaningful human control" and the impact thereof from the perspective of international law, including both the law of war and the law of arms control.

In the following, the argument sets off with a few remarks on the politics of terminology, then turns to time and place as challenges to the exercise of human control. Addressing cognition (and predictability) as well as volition (and intent) as elements of human control, the paper raises the question of value-based and discretionary decision-making. The contribution concludes by pointing out that instead of introducing new terminology the challenges of autonomous weapons and de-humanization of warfare can be better met by a contextual application of the law as it stands. The introduction of new concepts into an established debate only has limited benefits, if at all.

2 The Politics of Terminology

Whenever there is a need to stimulate debate, proponents raise the question how to attract the attention of the audience aimed at. Only if there is a minimum degree of attention by decision-makers, political and legal change can be initiated. Pertinent

[1]For a more general perspective on force protection under international humanitarian law see Neuman (2006) and Geiß (2012).

[2]In addition to the references below, the contributions by Beard (2013/14), Asaro (2012), and Noone and Noone (2015) are illustrative.

[3]Convention on Prohibitions or Restrictions on the Use of Certain Conventional Weapons Which May be Deemed to be Excessively Injurious or to Have Indiscriminate Effects, Geneva, 10 October 1980, UN Doc A/CONF.95/15, 27 October 1980, Annex I.

attention contributes to agenda setting, directing controversies into a particular direction, ensuring media coverage, and finally changing the balance of costs and benefits when turning to decision-making.

In the context of discussing the implications of de-humanized warfare, the "Campaign to Stop Killer Robots"[4] may serve as an example:

First, this campaign has managed to impact agenda setting in the context of the Fifth CCW Review Conference in 2016 when states decided to formalize their deliberations on lethal autonomous weapons systems,[5] after informal discussions had been held since 2014. Second, it has been successful to promote a critical attitude towards autonomous or semi-autonomous warfare by the mere use of the term "killer robots",[6] the image of which obviously does not only raise fears among the general public but also serves to delegitimize particular technological developments and their inclusion into governments' defence considerations. Third, albeit to a limited extent, the campaign has attracted a degree of media coverage—even though this has not reached the level of the campaigns to ban land mines and to eradicate cluster munitions.[7] Finally, states currently investing into weapons technology and automation will have to carefully consider the negative image of "killer robots" when taking pertinent budgetary or other decisions, irrespective of a more or less objective assessment of the pros and cons of any such decision on the agenda.

Terminology, however, is more than politics. Terminology matters in legal instruments:

Treaty texts normally reflect the importance of terminology by including an article or a section, which specifically deals with definitions. Such a provision reflects the particular relevance of terminology in the context of law making. Definitions included in such a provision normally are only meant to apply to the particular treaty. This is reflected in pertinent treaty language: "Use of terms" and "for the purposes of the present Convention" are typical phrases to this end. They indicate that terminology is the outcome of (sometimes long and protracted) treaty negotiations, that terminology reflects a compromise agreed upon between the negotiators and that such terminology is limited to the particular treaty.

In the context of the law of armed conflict Article 2 of the 1977 First Protocol Additional to the Geneva Conventions of 1949[8] (in the following: AP I) may serve as

[4]For information see https://www.stopkillerrobots.org. See also Human Rights Watch, Losing Humanity: The Case against Killer Robots (2012), available at http://www.hrw.org/sites/default/files/reports/arms1112ForUpload_0_0.pdf.

[5]Final Document (Advance Version), Fifth Review Conference of the High Contracting Parties to the Convention on Prohibitions or Restrictions on the Use of Certain Conventional Weapons Which May Be Deemed to Be Excessively Injurious or to Have Indiscriminate Effects, UN Doc CCW/CONF.V/10, 23 December 2016.

[6]The term is even taken up in academic journals, see, among others, Elias (2016).

[7]Wexler (2003), Bothe (2003).

[8]Protocol Additional to the Geneva Conventions of 12 August 1949, and relating to the Protection of Victims of International Armed Conflicts (Protocol I), 8 June 1977, 1125 UNTS 3.

an initial example. The heading is “definitions”, and the opening phrase of Article 2 reads “(f)or the purposes of this Protocol”. A second example in the context of (lethal) autonomous weapons can be taken from the various protocols attached to the CCW. Article 2 of Protocol II (on prohibitions or restrictions on the use of mines, booby-traps and other devices)[9] refers to “definitions” “for the purpose of this Protocol”. Similar approaches can be found in Article 1 of Protocol III[10] and Article 2 of Protocol V.[11]

What can be taken from this?

First, states do not use terminology in legal instruments for purposes of activism, not to say campaigning or propaganda; rather they use legal terminology for the purpose of delimiting and precisely defining the scope of legal obligations undertaken (often in view of preserving their sovereign rights). Second, pertinent provisions on terminology normally are the consequence or outcome of negotiations; they are hardly ever top-down and normally not pre-determined. Given this, the particular role of legal terminology must be sharply distinguished from the use of terminology in a predominantly political context.

It is against this background that the use of “meaningful human control” in the discussions about lethal autonomous weapons systems within the framework of the Convention on Certain Conventional Weapons must be assessed. It is worthwhile to recall the following:

At the 2013 CCW Meeting of High Contracting Parties, a new mandate on lethal autonomous weapons systems was agreed upon, stating that the “... Chairperson will convene in 2014 a four-day informal Meeting of Experts ... to discuss the questions related to emerging technologies in the area of lethal autonomous weapons systems. In the context of the objectives and purposes of the Convention.”[12] The report on this meeting[13] only once uses the term “meaningful human control”. Paragraph 20 of the report reads: “Many interventions stressed that the notion of meaningful human control could be useful to address the question of autonomy. Other delegations also stated that this concept requires further study in the context of the CCW.”

Another meeting of experts was held in 2015, based on the 2014 CCW Meeting of High Contracting Parties.[14] The annotated programme of work[15] twice refers to “meaningful human control”. It is understood as one of the elements “for an under-

---

[9]Protocol (II) on Prohibitions or Restrictions on the Use of Mines, Booby-Traps and Other Devices, Geneva, 10 October 1980, UN Doc A/CONF.95/15, 27 October 1980, Annex I.

[10]Protocol (III) on Prohibitions or Restrictions on the Use of Incendiary Weapons, Geneva, 10 October 1980, UN Doc A/CONF. 95/15, 27 October 1980, Annex I.

[11]Protocol (V) on Explosive Remnants of War, 28 November 2003, CCW/MSP/2003/2.

[12]Information available at http://www.unog.ch/80256EE600585943/(httpPages)/A038DEA1DA906F9DC1257DD90042E261?OpenDocument.

[13]CCW/MSP/2014/3 of 11 June 2014.

[14]CCW/MSP/2014/9, para. 36.

[15]CCW/MSP/2015/WP.1/Rev. 1.

standing on LAWS" and identified as a possible topic for discussion.[16] A working paper prepared for the meeting, labelled "food for thought", only once refers to the notion of meaningful human control. Under the heading of characteristics of LAWS, the following questions are laid down: "What is 'meaningful human control' of a weapon system? Does the level of human control assist in distinguishing LAWS from other weapons systems?".[17]

Another step forward was taken in 2016 with a third meeting of experts, based upon a mandate of the 2015 CCW Meeting of States Parties. In a letter of 17 March 2016, sent by the Chair, German Ambassador *Biontino,* with an attached "food for thought"-paper[18] the notion of "meaningful human control" is only referred to once, raising the question how "meaningful human control" of a weapon system can be defined. The advance version of the report of this meeting[19] more often refers to the concept. Paragraph 15 thereof reports that the concept was proposed "by some delegations as a framework to assess the legal, moral and ethical aspects of LAWS". The paragraph also states that, notwithstanding "broad interest in this concept", some participants pointed out that "there would be difficulties in identifying its scope". Others criticized the subjective nature of the concept and expressed a preference for "appropriate human judgement instead". It can be taken from further paragraphs in this report[20] that many perceived the concept as attractive while at the same time, however, there was criticism of its lack of precision. The chairperson of the informal meeting of experts submitted recommendations to the 2016 Review Conference, an advance version of which is available online.[21] It is interesting, that no reference was made to the concept of "meaningful human control" in these recommendations. The 2016 Review Conference decided to establish a Group of Governmental Experts on LAWS, which will meet in August and November 2017.

It can be taken from the above, that the concept of "meaningful human control" has served as a political driver in the discussions held in Geneva so far, but has not developed into legal terminology. This also seems to be a widely shared assessment among academics.

Thus, *Rebecca Crootof*[22] concludes that the concept is imprecise and can advance the discussion on the regulation of autonomous weapons systems only

[16]Ibid., paras. 5 and 6.

[17]CCW/MSP/2015/WP.2 of 20 March 2015.

[18]Document available at http://www.unog.ch/80256EDD006B8954/(httpAssets)/4423082AB7EF30E4C1257F7A00501844/$file/LAWSMX_FoodforThoughtFinal.pdf (Lethal Autonomous Weapons Systems, Informal Meeting of Experts, Geneva, 11–15 April 2016, Food-for-thought).

[19]Document available at http://www.unog.ch/80256EDD006B8954/(httpAssets)/DDC13B243BA863E6C1257FDB00380A88/$file/ReportLAWS_2016_AdvancedVersion.pdf (Report of the 2016 Informal Meeting of Experts on Lethal Autonomous Weapons Systems (LAWS)).

[20]Ibid., paras. 35, 38, 58, 68.

[21]Document available at http://www.unog.ch/80256EDD006B8954/(httpAssets)/6BB8A498B0A12A03C1257FDB00382863/$file/Recommendations_LAWS_2016_AdvancedVersion+(4+paras)+.pdf.

[22]Crootof (2016).

when it supports existing norms in the law of armed conflict. *Michael C. Horowitz* and *Paul Scharre*[23] in their paper on "meaningful human control" argue that further discussion and dialogue is needed to better understand the issues emerging from the notion of meaningful human control. Such issues according to their views are accountability, responsibility and controllability. I would conclude from their views that the notion of meaningful human control does not provide additional clarity beyond existing legal rules. A contribution by *Thompson Chengeta*[24] seems to be more positive on the notion, but does not go beyond accountability and responsibility, among others referring to criminal and command responsibility.

My conclusion from these assessments is that the concept of "meaningful human control" should not be framed and used as a legal concept. Rather, use should be made of existing and (largely) undisputed rules of international humanitarian law and possibly of arms control law in order to define the limits of lethal autonomous weapons systems.

In the following three sub-sections I will identify some of the issues that have already been taken up in the context of this debate, demonstrating that the concept rather is a political label bringing together various legal issues that are of importance in the context of lethal autonomous weapon systems. It does not carry the potential to develop into a legal principle nor a rule impacting the behaviour of relevant actors in the law of armed conflict or arms control law.

3 Time and Place as Challenges of Targeting

Time and place of decision-making, of action and of effects may always differ, but they certainly do in situations of armed conflict. In order to protect themselves humans have long ago made use of distance weapons. Far beyond this, modern technology has facilitated the development of weapons, which allow many different combinations of spatial and temporal factors. Missile technology has helped to reduce time between the actual launching and the impact on the target. Unmanned aerial vehicles, on the other hand, can stay close to a target, controlled at nearly any kind of distance, for hours if not days. Remote piloting of such vehicles allows better reconnaissance and may enhance the effectiveness of military operations.

The question arises whether these enhanced options to combine time and place of decision-making, action, and effects, are adequately addressed by the law as it stands, or whether the notion of "meaningful human control" is indispensable to fill in potential gaps.

To this end, it is important to distinguish between the legality of particular types of weapons *per se* and the unlawful use of weapons.

---

[23] Horowitz and Scharre (2015).

[24] Chengeta (2016).

There are only very few rules prohibiting weapons *per se*. Among them is Article 35, paragraph 2, AP I,[25] which protects combatants and prohibits the employment of "weapons, projectiles and material and methods of warfare of a nature to cause superfluous injury or unnecessary suffering." Leaving aside that lethal autonomous weapons systems are not weapons as such but consist of "a weapon and the items associated with its employment",[26] it is possible that, depending upon the circumstances and upon the effects on the targeted individual, lethal autonomous weapons systems violate Article 35, paragraph 2, AP I—but this is only a possibility, and as such cannot be considered "a valid basis for imposing an across-the-board pre-emptive ban on the systems".[27] The concept of "meaningful human control" does not add anything at this point since Article 35, paragraph 2, AP I does not address the manner of engagement but focuses on the effects caused by a particular weapon.

As far as the unlawful use of weapons is concerned, it may suffice—in light of other contributions in this volume—to take Article 52 AP I as an example. Its paragraph 1 stipulates: "Civilian objects shall not be the object of attack or of reprisals. Civilian objects are all objects which are not military objectives as defined in paragraph 2." The definition of military objectives is included in paragraph 2: "Attacks shall be limited strictly to military objectives. In so far as objects are concerned, military objectives are limited to those objects which by their nature, location, purpose or use make an effective contribution to military action and whose total or partial destruction, capture or neutralization, in the circumstances ruling at the time, offers a definite military advantage." Paragraph 3 adds a legal presumption: "In case of doubt whether an object which is normally dedicated to civilian purposes, such as a place of worship, a house or other dwelling or a school, is being used to make an effective contribution to military action, it shall be presumed not to be so used." The term "attack" is defined in Article 49 AP I as follows: " 'Attacks' means acts of violence against the adversary, whether in offence or in defence."

Article 52 AP I is understood as one of the most important provisions on the legal frame for targeting.[28] While it addresses the (expected and envisaged) effects by reference to "total or partial destruction, capture or neutralization" and to "definite military advantage", there is little on the actual action, or the execution of an attack, except that Article 49 AP I describes these as "acts of violence". The primary focus of Article 52 AP I is on the selection of targets, their prioritization, and the planning of the actual attack. In other words, Article 52 focuses on the time preceding the attack. It is inherent in this provision, that there is a time difference between decision-making, action, and effects. The provision assumes that differences in time and place between decision-making, action, and effects are of limited relevance, as

[25]Boothby (2012), pp. 258–260; Boothby (2016), pp. 50–55.

[26]Schmitt (2013), p. 3.

[27]Ibid., p. 9.

[28]Schmitt and Widmar (2016), pp. 129–135.

long as decisions are taken in light of envisaged action and effects. In as much as Article 52 AP I requires the impact on the target to be taken into account at the time of decision-making in respect of an attack, again the concept of "meaningful human control" does not add anything beyond this existing legal rule.

In addition, from a practical perspective and for the time being (as *Michael Schmitt* has pointed out), "a fully autonomous system is never completely human-free. Either the system designer or an operator would at least have to program it to function pursuant to specified parameters, and an operator would have to decide to employ it in a particular battlespace."[29] Full autonomy thus refers to "the capability to identify, target, and attack a person or object without human interface".[30] Proceeding from this, differences in time and space do not require any addition to what is included in Article 52 AP I. There is no need to refer to "meaningful human control" since this will not add more precise criteria to the requirements of Article 52 AP I, which an autonomous system must meet. Rather the opposite must be borne in mind: the criteria included in Article 52 AP I are much more precise than those emerging from "meaningful human control".

4 Cognition (Predictability) and Volition (Intent) as Elements of Human Control

Based upon the above considerations, cognition and volition likewise are matters of decision-making. However, they establish a link between decision-making, action and effects. Cognition refers to knowledge of the decision-maker about the consequences of particular decisions and acts. Causality between decision and action and between action and effects is the most pertinent linkage in so far. However, knowledge is not limited to the ordinary or natural chain of cause and effect (*conditio sine qua non*) but it must also bear in mind unexpected side effects and exceptional chains of causation with their respective probability.

It is not unlikely that target recognition technology of autonomous weapons systems actually is better able to predict the effects of a particular use of weapons.[31] In order to make this possible those who design weapons systems will have to ensure that they are in a position to recognize objects or persons that are lawful objects of attack. This means, in an international armed conflict, to only attack military objectives according to Article 52 AP I (not civilian objects), and combatants (not civilians according to Article 50 AP I, bearing in mind Article 51 AP I), also keeping an eye on appropriate rules of customary international law. In a non-international armed conflict, Articles 13–18 AP II,[32] Common Article 3 of the four Geneva

[29] Schmitt (2013), p. 4.

[30] Schmitt and Thurnher (2013), p. 235.

[31] Sassóli (2014), pp. 310–311.

[32] Protocol Additional to the Geneva Conventions of 12 August 1949, and relating to the Protection of Victims of Non-International Armed Conflicts (Protocol II), 8 June 1977, 1125 UNTS 609.

Conventions, and pertinent rues of customary international law have to be borne in mind. While the application of the principle of distinction may be difficult in respect of human beings,[33] it is most likely that autonomous recognition of items of military hardware will not only be realistically possible using currently available technology. Even more, it may often be much more reliable than human involvement.

It is useful at this point to recall the so-called *Vincennes* incident. In 1988, during the Iran-Iraq war, an Iranian Airbus with civilian passengers was downed by mistake by the US cruiser *Vincennes* in the Persian Gulf. The airliner ascended on its usual climb profile. It was correctly picked up by the cruiser's radar. However, the commander of the *Vincennes* erroneously considered it to be an Iranian F-14 with hostile intent and ordered it to be downed. Subsequent investigations by the US and the ICAO attributed the downing to "a combination of human error and technical shortcomings".[34] Civilian and military frequencies for broadcasting were not linked to each other, and the airliner's silence supported the commander's belief in an imminent threat. It was argued by the US that the crew had followed a trained pattern in light of a previous Iraqi missile attack on another US vessel and the narrow time window for a decision, disregarding data pointing to the contrary. This underscores that the notion of "human control" does not really add anything to the cognitive aspect.

In the application of Article 57 (2) (a) (ii) AP I, which requires parties to the conflict to do everything feasible to verify that the object of the attack is a military objective,[35] it is also highly likely that autonomous weapons systems (i.e. algorithm-based technology) are in an equally good, if not better position to differentiate between the objects they are programmed to identify and those they are not, i.e. between military objects and civilian objects. The same quality does not seem to be available with a view to identifying, and attacking, human targets, under this first element of Article 57(2)(a)(i) AP I.

Human operators may, however, bring in some added value to the cognitive aspect, namely the understanding of more abstract circumstances, such as the knowledge that there is an armed conflict, or that a particular activity is part of an overall plan, or others. But this does not require actual and "meaningful human control"; it only underlines the overall benefits of a degree of human involvement.

As far as volition is concerned, it is generally understood that an autonomous system cannot have intention, even though *Gabriel Hallevy* proposes that if a system is programmed with a purpose and it subsequently acts in furtherance of this purpose the system acts with intent.[36] However, most seem to support the

[33]This may be the background for the current U.S. policy which provides: "Human-supervised autonomous weapon systems may be used to select and engage targets, with the exception of selecting humans as targets ..." (U.S. DoD Directive 3000.09: Autonomous Weapons Systems: Response-To-Query Talking Points, § 4c).

[34]Hailbronner and Hellmann (2009), para. 15.

[35]Gillard (2016), pp. 174–176.

[36]Hallevy (2010a), p. 189; Hallevy (2010b), p. 18.

view that an autonomous system is incapable of forming intent, which is where human control comes in. Due to the fact that it is bound by its programming parameters and even in the case of 'strong' autonomous intelligence, it would be contentious to state that an autonomous system has true free will, especially from a philosophical perspective and without free will there can be no intent. Thus, the intent follows from human involvement and remains there, either through supervision, programming or deployment. Again, however, this illustrates the place of volition, but does not necessarily require the notion of "meaningful human control" in legal terms.

Furthermore, the notions of cognition and volition are the focus of personal criminal accountability (*mens rea*) more than of human control and concern therefore more the accountability of the human who has control.

5 Value-Based and Discretionary Decision-Making?

While the above Sects. 2 and 3 illustrate that the notion of "meaningful human control" does not necessarily add anything to the law as it stands, this may be different when it comes to value-based and discretionary decision-making as it is included, among others, in Article 57(2) (iii) and 57(3) AP I. One might, in particular, consider discretionary decision-making as the basis for distinguishing autonomous and automated weapons systems.[37] However, the criteria for decision-making arise from the provision itself; they do not arise from the concept of "meaningful human control". Thus, there is virtually nothing that may derived from this concept beyond what can already be drawn from the existing rules on targeting.

Even more, it must be borne in mind that, due to sensor technology and algorithms, it may be possible for a system to at least identify civilian objects, perhaps even civilians. However, there is a problem with assigning objective and absolute value either to these objects as well as to the military advantage. Currently, proportionality decisions are already taken with the help of computer programs, but the final decision on specific attacks is still made by humans.[38] This illustrates that here the involvement of a human seems to be necessary, as a decision on proportionality is not completely objective, and the specific nuances cannot easily be programmed into a system, as each and every single object would be needed to be given a pre-set value. Again this emerges from the very notion of proportionality, and not from the concept of "meaningful human control", which only serves to describe or highlight matters.

It may thus be argued that, in this regard again, human control, rather than being a legal requirement in itself, may be perceived as an approach to ensure compliance and implementation of an existing legal requirement, i.e. the law of armed conflict.

[37] On the distinction see Wagner (2016), paras. 1–6.

[38] Arendt (2014), pp. 30–34.

Indeed, discretionary decision-making, in the context of meaningful human control, appears to focus on the critical functions in the targeting process, where a human is involved to ensure that the use of force complies with the law of armed conflict. Meaningful human control in this regard therefore emerges primarily as a standard of implementation aiming at the proper application of the law of armed conflict.

6 Meeting the Challenges of Lethal Autonomous Weapons Systems by Contextual Application of the Law

From the perspective of the author, the challenges of lethal autonomous weapons systems can better be met by the contextual application of the law as it stands than by developing new law. The concept of "meaningful human control" facilitates implementation of the law as it stands but it does not add anything new to existing standards. Rather the inclusion of this concept runs the risk that it blurs some of the clarity included in existing rules. While it has the potential to become a buzzword, it carries the risk to weaken existing law.

The law thus should not be changed, and the concept of meaningful human control should not become part of the law. It may, however, serve other purposes if it is kept to the political and non-legal sphere.

As explained in the introductory section, terminology matters. It may serve different purposes, but in the context of international law it facilitates the precise determination of rights and obligations. "Meaningful human control" is not a concept suitable for such clear-cut specifications. It draws attention to relevant considerations, but if inserted into the law it would most likely weaken international norms rather than strengthening them. The concept thus is useful in the political sphere and should be avoided in legal terms.

References

Arendt R (2014) Der Einsatz autonomer Waffensysteme im Lichte des Verhältnismäßigkeits- und des Unterscheidungsgrundsatzes. In: Frau R (ed) Drohnen und das Recht. Mohr Siebeck, Tübingen, pp 19–34

Asaro P (2012) On banning autonomous weapon systems. ICRC Int Rev Red Cross 94:687–709

Beard JM (2013/14) Autonomous weapons and human responsibilities. Georgetown J Int Law 45:617–681

Boothby WH (2012) The law of targeting. Oxford University Press, Oxford

Boothby WH (2016) Weapons and the law of armed conflict. Oxford University Press, Oxford

Bothe M (2003) Durchsetzung der Menschenrechte und des humanitären Völkerrechts – ein Paradigmenwechsel? In: Marauhn T (ed) Die Rechtsstellung des Menschen im Völkerrecht. Entwicklungen und Perspektiven. Mohr Siebeck, Tübingen, pp 115–130

Chengeta T (2016) Defining the emerging notion of 'Meaningful Human Control' in autonomous weapon systems (AWS). Available at https://ssrn.com/abstract=2754995

Crootof R (2016) A meaningful floor for "Meaningful Human Control". Temple Int Comp Law J 30:53–62
Elias RA (2016) Facing the Brave New World of Killer Robots. Indones J Int Comp Law 3:101–126
Geiß R (2012) The principle of proportionality. 'Force Protection' as a military advantage. Israel Law Rev 45:71–89
Gillard E-C (2016) Protection of civilians in the conduct of hostilities. In: Liivoja R, McCormack T (eds) Routledge handbook of the law of armed conflict. Routledge, London, pp 157–180
Hailbronner K, Hellmann D (2009) Aerial incident Cases before international courts and tribunals. In: Wolfrum R (ed) The Max Planck Encyclopedia of Public International Law. Oxford University Press, Oxford, Online Edition (www.mpepil.com)
Hallevy G (2010a) The criminal liability of artificial intelligence entities – from science fiction to legal social control. Akron Intellect Prop J 4: 171–201
Hallevy G (2010b) Virtual criminal responsibility. Orig Law Rev 6:6–27
Horowitz MC, Scharre P (2015) Meaningful Human Control in Weapon Systems: A Primer, CNAS (Centre for a New American Security) Working Paper, available at https://www.cnas.org/publications/reports/meaningful-human-control-in-weapon-systems-a-primer
Neuman N (2006) Applying the rule of proportionality: force protection and cumulative assessment in international law and morality. Yearb Int Humanit Law 7.2004:79–112
Noone GP, Noone DC (2015) The debate over autonomous weapons systems. Case West Reserve J Int Law 47:25–35
Sassóli M (2014) Autonomous weapons and international humanitarian law: advantages, open technical questions and legal issues to be clarified. Int Law Stud 90:308–340
Schmitt MN (2013) Autonomous weapon systems and international humanitarian law: a reply to the critics. Harv National Secur J. Features 05 February 2013, 1–37. Available at http://harvardnsj.org/2013/02/autonomous-weapon-systems-and-international-humanitarian-law-a-reply-to-the-critics/
Schmitt MN, Thurnher JS (2013) 'Out of the loop': autonomous weapon systems and the law of armed conflict. Harv Nat Secur J 4:231–281
Schmitt MN, Widmar E (2016) The law of targeting. In: Ducheine PAL, Schmitt MN, Osinga F (eds) Targeting: the challenges of modern warfare. T.M.C. Asser Press, The Hague, pp 121–146
Wagner M (2016) Autonomous weapon systems. In: Wolfrum R (ed) The Max Planck Encyclopedia of Public International Law. Oxford University Press, Oxford, online Edition (www.mpepil.com)
Wexler L (2003) The international development of shame, second-best responses, and norm entrepreneurship – the campaign to ban landmines and the Landmine Ban Treaty. Ariz J Int Comp Law 20:561–606

Concluding Remarks by the Editors

Wolff Heintschel von Heinegg, Robert Frau, and Tassilo Singer

Abstract This chapter summarizes the problems and solutions of dehumanization of warfare. As will be shown, legal aspects are only part of a wider array of problems, especially technical and ethical questions. However, existing law is generally well suited to regulate the conduct of sophisticated weaponry. Adjustments are needed in some parts of the law.

The challenges of the development and use of technology which enables the conduct of hostilities without human control proves to be manifold and complex.

Both autonomous weapon systems and cyber operations are thought to be able to conduct attacks without human control in the future, which means that the term "dehumanization" can be used as a generic term.

As human decision making is integrated by the requirements of core rules of the so called 'targeting law', it can be confirmed that the replacement of humans out of decision making processes is relevant from a legal point of view. Therefore, the question has to be answered whether and to what extent dehumanization of warfare is in compliance with LOAC.

The Dehumanization of Warfare has to be perceived as multi-layered. The reduction of human control is closely linked to autonomy of unmanned systems and computer control software. However, the notion of autonomy is already controversial itself. A common understanding of the term, its scope and its subcategories is, therefore, imperative and would foster the ongoing discussion about autonomous weapon systems. Artificial intelligence as a notion might add to the discussion in the future, because it is closely linked to the understanding of a high degree of autonomy. In addition, a clear distinction between a system which allows human control and decision overriding, and a system without these characteristics is needed with a view to their lawfulness.

W. Heintschel von Heinegg (✉) • R. Frau
Faculty of Law, European University Viadrina, Frankfurt (Oder), Germany
e-mail: heintschelvonheinegg@europa-uni.de; frau@europa-uni.de

T. Singer
University of Passau, Passau, Germany
e-mail: tassilo.singer@uni-passau.de

W. Heintschel von Heinegg et al. (eds.), *Dehumanization of Warfare*,
DOI 10.1007/978-3-319-67266-3_12

A key challenge for state of the art robots, which are to perform their intended duties autonomously, is that they have to cope with real life environment. This means that they have to adapt their behaviour to every new situation and to possibly changing goals during the execution of their mission. From the technical perspective autonomous robots (e.g. military robots in the future) and automatic robots (e.g. industrial robots) must be distinguished. This decision depends on whether the robot is used in an open environment (more difficult) or a closed environment. Autonomy, therefore, should be defined generally as "freedom from external restraint and self-government" and "decisional autonomy". However, a more precise definition of autonomy depends on the context. The decisive point is the decision-making of autonomous robots, meaning the reaction to unknown environment, unforeseen situations and behaviour of other objects and subjects in the operation area. The behaviour of a robot as an intelligent (rational) agent is framed by certain parameters like the performance measurement and cost function, the available actions, the knowledge about the environment and the sensor information and finally the system internal conclusions. Still these functions and abilities could be qualified as automatic. An autonomous system will furthermore require (machine-) learning abilities. Currently available algorithms are not yet robust in complex environments and all possible situations. Even if the sensors, the sensor fusion, the modelling and the interpretation would work perfectly, the execution might still differ from the intended result. Therefore, from the perspective of current state of the art, a human should still always oversee the actions of a robot.

Autonomous actions or behaviour can be a part of the tools used for cyber operations. For example, the worm StuxNet operated autonomously after its insertion into its target-systems. Past cyber operations were used either for very limited operations, not reaching the threshold of an "armed attack" by themselves or as a part of a whole war campaign (as in Georgia 2008). A war solely based on cyber methods and means of warfare has not taken place so far. Regularly, cyber operations form part of a war campaign entailing all types of forces and will not be conducted as standalone, but it cannot be excluded that a sole cyber war might happen in the future. In case a cyber attack takes place which reaches the required threshold, the analysis of the operation and its origin are of greatest importance for possible countermeasures. However, a clear and irrebuttable attribution poses technical difficulties, which result in limited possible reactions. These issues can be addressed and partly solved by digital forensics.

The main function of digital forensic is to collect digital traces and to reconstruct digital evidence for the attribution process. The process of (scientific) digital forensics can be subdivided in identification, individualization, association and reconstruction. For the forensics process hypothesis, testing is a crucial element. Digital evidence has to be obtained via an analysis of multiple layers of content. It can be persistent, semi-persistent and volatile. However, digital evidence can be manipulated or deleted and thereby aggravate any attribution or even render it impossible. Therefore, authenticity, the probative value and the integrity of evidence has to be checked and categorized in levels of certainty. Nonetheless, attribution based solely on digital evidence is difficult. For example, in order to attribute a

DDOS attack, regular internet traffic has to be distinguished from malicious traffic leading to the source of the attack. Even if the original IP address can be acquired, it might have been manipulated. Thus, additional sources would be needed for clear and convincing evidence.

The law itself does not pose a direct challenge on the first level, because it is and has to be applicable to warfare with or without human control as is argued in several contributions of this volume. In addition, the general principles of the law of armed conflict are applicable. Thus, the rules for autonomous weapon systems and cyber operations are the same as in conventional warfare.

However, on a second level, the common understanding and interpretations of some rules might not be directly transferrable to new forms of warfare such as autonomous weapon systems, but have to be adapted to comply with the purpose of the law.

First of all, the legal classification of unmanned systems is very important.

Generally, the distinction between platform—meaning the weapon system—and the payload or weapon is of relevance for the lawfulness of each type. The legal classification also has to encompass the differentiation of the autonomous abilities. As previously mentioned, the factor of human controllability is decisive from a legal point of view. Thus, systems which can be controlled by a human—even if this is limited to intervention and overriding commands—have to be viewed as semi-autonomous. Because there is a way for human interference, such systems are, from a legal perspective, not different from common and conventional weapon (system) use, which is always controlled by humans. Thus, such systems and their use generally have to be considered as in accordance with current law. The possible legal challenge, therefore, can only consist of fully-autonomous weapon systems and cyber tools with the "man out of the loop". Both means of warfare must comply with the same legal framework.

The test for compliance of means of warfare with the law is incorporated in Art. 36 AP I and customary international law as an obligatory legal review. This is of great relevance for new means of warfare like cyber weapons and autonomous weapon systems. States, which are party to the AP I, additionally have to review new methods of warfare.

Regularly, only weapons law, which would prohibit a weapon per se, would have to be considered in the review. With a view to current unmanned systems, remote attacks and personnel distance do not pose a legal challenge except for the need for security measures in case of absence of human control.

In case human control is not available during the targeting process due to the autonomous self-guidance design of a system or tool, the compliance of the system with targeting law has to be tested in the review, too.

The review of means of warfare in the weapons law implies the legality of weapons and systems per se. Thus, if a system cannot comply with these rules, it cannot be used at all. The most relevant rules for the dehumanization of warfare are the prohibition of weapons of a nature to cause superfluous injuries or unnecessary suffering and the prohibition to use weapons of an indiscriminate nature.

Depending on the technical abilities of unmanned systems and tools of cyber warfare, these rules do not pose an unsolvable challenge to systems with artificial intelligence and sufficient sensor data processing in the long term. Thus, in general, weapons without human control (dehumanized weapons) such as LAWS are not prohibited per se. This is the case especially for Art. 35 (2) AP I, which is solely focused on the effect of weapons and not on the type of control of the weapons.

Secondly, fully autonomous systems have to comply with the operational law and especially the law of targeting. Core rules of the law of targeting are the principle of distinction, the duty to precautions and the principle of proportionality. Again, if the decision making is transferred to a system or program with autonomous abilities and no link for human overriding exists, the system itself has to act in conformity with the law. Therefore, it becomes a question of technology and technical abilities, which either enable compliant and lawful behaviour or not. This again also depends on the preciseness of the review and the perspective of the reviewer, as there are no common and internationally accepted standards of procedure for the review. If the thresholds are interpreted widely, one reviewer could regard a system lawful whereas others would not. Furthermore, the environment of the use is important and the key factor for determining legality.

It is submitted that compliance with the principle of distinction is possible if the system or tool has sophisticated technological abilities. The proportionality rule requires the ability to evaluate military advantage, to determine possible incidental damages and injuries and to compare and weigh these legal factors with each other before reaching a decision. It is at least required to have such sophisticated abilities for understanding the context of an operation and complex scenarios. Also for values cannot be established by a system from the state of the art of current technology. Therefore, the possibility of overriding decisions of an autonomous weapon system by a human is obligatory.

However, this does not preclude a use in general, because missions with an "easier" environment could still be conducted without the necessity to comply with the proportionality rule. It can be argued that in a defence situation a different use would be given. If the defence system can distinguish its intended target and other, protected targets, autonomous weapon systems could also be in conformity with the proportionality rule. This argument can be extended to a use in sparsely populated areas which, it is submitted, is lawful depending on whether the (positive) assessment is made by a human in advance. It is thus stressed that if decision makers take the necessary precautions, the legal hurdles could be overcome.

The same holds true for the precautions rule, which also requires value based decision making. The hurdles there can be circumvented by restrictions to certain operations, too. A solution would be to include (obligatory) the possibility to cancel or suspend an attack in cases where collateral damage becomes evident during attack. The feasibility rule may require the use of a manned or man-controlled system, if an autonomous system cannot comply with Art. 57 (2) (2) (i) and (ii) AP I. Art. 57 (2) (a) (i) AP I poses a challenge for autonomous weapon systems in that it requires a separation between lawful targets like able-bodied combatants and unlawful targets. Also an autonomous weapon system has to be able to verify that it is not prohibited to attack a specific target by law. Because this question depends on future technology, it has, for now, to be left unresolved.

Warnings as a precaution may be easier using autonomous systems, too, because these do not have the same need for self-preservation as humans. It might be challenging, though, to implement a good way to select between different targets in a system. Other problems are posed by the need of distinguishing a use of AI robots for military and/or civilian purposes from an active and passive perspective. Another challenge lies in the inherent unpredictability of the behaviour of fully autonomous systems.

In the overall review on LOAC it has to be stressed that no rule seems to require specifically human control or that only a human can comply with a rule. Rather the rules set (high) standards of behaviour, with which everyone and everything participating in hostilities has to comply. Difficulties by limited abilities of a system could be countered by limiting operation parameters. From an ethical point of view the perspective of possibly lawful autonomous weapon systems in the future might be viewed differently, as discussed later.

The current technology just cannot fulfill all of these legal thresholds right now. Consequently, there has to be a minimum degree of human control for the use of autonomous systems to fulfill all legal prerequisites. In contrast, for a minimum use in a restricted environment fully autonomous systems have to have at least the ability to comply with the general principles of the law of armed conflict.

Other questions of this volume were related to specific legal issues like the status of unmanned maritime systems. At least these can be qualified as belonging to armed forces of a State. The rules of weapons law and targeting applies to unmanned maritime systems in armed conflicts at sea, too, as does the law of neutrality with navigational rights for neutral States.

Acts of cyber warfare also have to comply with the LOAC. Cyberspace owns big vulnerabilities. It is mostly based on common platforms, software and protocols and it is inherently intertwined. Any manipulation can be followed by an uncontrolled proliferation and result in cascade failures. Public life and society can be harmed severely, which confirms the comparability of cyber warfare to WMDs.

Cyber weapons can also be considered as weapons if they are designed or intended to cause violent consequences. As cyber weapons are regularly designed for a specific mission or attack, they should not be legally reviewed in advance. Rather the review should take place (shortly) before the mission by a legal advisor of the operational commander, who is familiar with the law for cyber warfare. The review of a cyber weapon has to consider the specific circumstances of the attack and check the relevant law accordingly. For the review, all different levels of orders and effects of cyber weapons have to be taken into account. This includes the impact on the performance of the computer system, the effect on the facility wherein the computer is used and the damage or/and destruction suffered. With a view to the possible effects of the tools, also (weapon-) specific rules like the prohibition of booby traps have to be included. The definitions of these particular weapon prohibition treaties have to be considered and transferred carefully in the legal review. If cyber weapons are designed and used for tampering of other weapons and systems, the relevant rules for these weapons have to be observed.

Some means of cyber warfare can violate the prohibition of indiscriminate attacks if their attacks cannot be limited to lawful targets. The rules for lawful targets such as military objects can be transferred to cyber warfare. It is open to discussion whether the threshold for the military advantage of an attack on a random target (with a view to uncontrollable cyber worms) has to be high. In the context of the proportionality rule it is required that the military advantage is concrete and direct, which is interpreted by some as a stricter standard than a "definite" military advantage. Cyber attacks against dual-use objects might become problematic because a rather strict proportionality test has to be applied. However, if the gravity of possible impacts is counted in the proportionality assessment, the rule might not only hinder the use of cyber attacks. It could also require a cyber attack instead of a conventional attack, if e.g. the bombing of infrastructure causes heavier effects and graver damage than a comparable cyber operation.

The discussion of the term (meaningful) human control has shown that terms can be, and often are, exploited for political ends. It can be criticized that often other groups than the original legislators in international law—the States—try to push the debates about weapons forward in their own interest. This is built upon the quest of seeking pertinent attention and agenda setting. However, this often results from misperceptions of real and unreal threats of these tools of warfare for society and mankind in general. The term meaningful human control, for example, has been brought up by the "Campaign to Stop Killer Robots", which thereby was able to promote a critical attitude towards LAWS. Consequently, the campaign set the agenda for the CCW Review processes and was able to affect the legitimacy of the process to develop laws and comparable technology by States.

In the relation between rules and the notion of meaningful human control the unlawful use of weapon systems and the targeting process, Art. 52 AP I becomes relevant. This provision is directed at the time before an attack. Inherently the time differs between decision-making, the act and the impact(s) and thus is only of limited relevance if decisions are taken in light of the envisaged action and effects. Hence, it is concluded that Art 52 AP I obliges to include the impact in the decision making process either way, with or without human control. The criteria deriving from the standard of meaningful human control therefore are seen as less precise than the criteria contained in Art. 52 AP I. It is not excluded that future target recognition technique can be superior to human abilities. Particularly the recognition of military hardware by autonomous weapon system is said to be more reliable than human involvement. Hence, human control is not seen as an additional benefit.

The equal or even superior abilities of autonomous weapon systems for object recognition are also important for compliance with the rule for precautions in advance of an attack. However, the identification and attack of human targets seems to be more complicated than with mere objects. The advantage of human operators thereby lies in the capacity to understand and consider complex situations and link unrelated circumstances. The concept of (meaningful) human control stresses the benefits of a certain level of human involvement. Additionally, volition proves to be problematic for autonomous weapon systems, because they are not able to have intent and (consequently) a true free will. The intent is depending on the human involvement by supervision, programming and deployment. The identification of targets through algorithms and systems could be technically feasible, but the value assignment to targets and the military advantage are problematic. Human control can foster compliance with these legal requirements of proportionality. However (meaningful) human control is not a condition for the conformity with LOAC but should be perceived as a standard of implementation.

Some classify the term as a framework for the legal, moral and ethical aspects, others criticize that the scope is hard to determine and that the concept is of subjective nature and lacks precision. It can be summarized that the notion of meaningful human control does not additionally clarify beyond existing law and that it should not be seen and relied upon as a legal concept. Instead, a contextual application of the existing law should be preferred. Thereby, the challenges of autonomous weapon systems and dehumanization could be met instead of trying to invent new rules and terms of law, which are imprecise. To support a demand for a total ban of such system would need a clear legal basis for the term or a change of the existing law. But this again would bear the risk of blurring the accuracy of the existing law. Thus, it can be concluded that the notion of meaningful human control should better be understood as a driver to advance political discussions, especially in the CCW process.

It has to be pointed out, that humans will always be involved at a certain point of the development and the use of an autonomous weapon system. The controlling software has to be programmed and every autonomous system needs to be activated at some point. Also a system always has to have parameters of the actual mission inserted including the waypoints, the area of operation and the duration of the mission. Therefore, the role of the military commander has to be underlined. Thus, the use of autonomous systems is also challenging with view to international criminal law, especially concerning the question of humans being held criminally liable and/or accountable for misbehaviour of AI robots. But, the necessary involvement of a human at some point of the use has to be connected with this question.

But even before such a system is used it has to pass the legal review test conducted by a human. This is at least the latest point to prevent an autonomous weapon system that is incapable of acting in accordance/in conformity with the law. If the system can only act in set parameters the common understanding of the law and its rules can be applied. But there might also be systems which have the ability to progress, e.g. to machine learning or to leave set parameters, and their actions are left totally to the internal situation assessment. It is held that such a system has to be regarded differently in legal terms. This applies as long as the "self-initiated" decision does not lead to a self-restriction of permitted acts which would be in accordance with the law (because then there is no legal need to act). From another legal perspective the critical issue might be the tactical attack decision. Thus, an AI system that can override preset constraints is considered to be unlawful.

Thus, it can be held the interpretation and understanding of the discussed law can and has to be adapted accordingly.

Some specific questions and challenges however remain unresolved: It is required by Art. 41 AP I and according customary international law that soldiers hors de combat have to be recognized and, accordingly, also by a fully autonomous weapon system. Is this realizable and based on which criteria? Is the threshold there higher due to superior sensor abilities or equally to the required recognition by a human? What is the legal status of data, especially concerning the criteria for a military object or a dual use object? The equipment with means of warfare is also a question that States need to address. What is an unmanned maritime (or naval) system? A mine, or a torpedo have to be excluded as they are not systems. Nevertheless unmanned naval systems have a wide range of targets which could influence the legal classification. Can an autonomous weapon system be classified as a ship, a vessel or a warship? Accordingly sovereign immunity and certain navigational rights could be encompassed. Can all the linked rights and duties of the law of the sea and the inherent rights of state ships like transit passage be transferred on unmanned (and autonomous) systems? What is the legal frame-work governing the tampering of unmanned systems, which becomes highly relevant concerning fully autonomous weapon systems? How can changes and damage to data be legally qualified? Can components of decision making processes and appreciation of values and their weighing be transferred into a logical system which might be inserted? What would be sufficient criteria? How is the collateral damage of dual use data or servers that are used for civilian and military purposes determined? Regularly they would be viewed as military objects, however, if this rule is strictly transferred and observed it would lead to a classification of most of the internet as a military object and thereby targetable. This list is just exemplifying and could be continued.

At the origin of IHL also ethical considerations can be found. The dehumanization raises several ethical questions especially concerned with the possibility that a dehumanized system could kill a human without any chance for a human to override or intervene against the systems decision.

The outcome that a war is justified ethically does not automatically allow to use all sorts of methods of warfare. The dehumanization understood as the removal and distance of the human soldier from the battlefield cannot be seen unethical per se with view to the duty of a State also to protect its own soldiers as well as possible.

The outsourcing of the kill decision to an autonomous weapon system could be viewed critically depending on an appropriate ethical criterion. The insidiousness of autonomous systems like an unknown danger could be invoked, however in a known state of armed conflict this criterion is not convincing as surprise in military operations is not considered damnable per se. Also, no prohibition can be based on the principle of distinction per se, if a proper distinction between military and civilian targets can be performed by autonomous weapon systems. Nevertheless it has to be analysed from an ethical standpoint, if an autonomous weapon system is able make such a (kill-) decision at all. The decisive point from an ethical perspective is, that a system—even with high autonomous abilities—may have cognitive abilities. However, in the view of Joerden it cannot evaluate a situation properly and perform normative decisions, which is however required for certain rules of LOAC. Based on this finding autonomous weapon systems can be objected ethically with view to their inability to oblige the law of armed conflicts in its entirety.

Linked to this question the ethical admissibility may differ from the admissibility in LOAC. From an ethical perspective evaluation rules (judgment about the behaviour as obligatory, forbidden or permitted) and imputation rules (attribution of act to a person and the responsibility of the person) have to be distinguished. Before one act is unethically, one must be able to attribute the specific act to a person and only then it can be evaluated if this act was illegal. Even if the action in question was unlawful under IHL, there may not be criminal accountability. Because of the rules of attribution or imputation, individuals may "unjustifiably" evade their own criminal responsibility.

Other challenges will be posed by a related and partly intertwined development: the enhancement of the human by technology. By this adaption of humans and their abilities with technology and software common legal problems and current problems with software and artificial intelligence become combined. First, there needs to be a differentiated approach to different forms of enhancement. Each of those has different legal problems. It is common that an enhanced human still has to be considered a human being and not a weapon or weapon system. The prohibition to cause superfluous injuries is problematic with enhancements. For example does it constitute a superfluous injury or unnecessary harm if an integrated chip has an electrical overload? In case of enhanced POWs the prohibition to experiment on POWs is specifically highlighted. Besides LOAC a major legal issue are human rights of enhanced individuals. Human rights law applies to them because they are humans. Also there is a strong need to look at the return of an enhanced soldier into society. With regard to responsibility, criminal law and state responsibility apply. But there are challenges to the mens rea of alleged perpetrators. Still questions remain open.

For example an unlawful attack happened, which caused excessive collateral damage. It was ordered or conducted by a human modified with microchip-led sensors, which developed and transferred the decisive information about the target situation to the human. In this case you could raise the question if the human acted or the machine. Who would bear the responsibility and is it the sole responsibility of the human? What about the liability of producers or software programmers? Can an enhanced human loose its status at some stage of transgression to a machine? And if so, what would be relevant parameters and factors of a human to be (considered) a human? Finally also from an ethical perspective you could ask if and how the artificial change—some might call it improvement—of a human is morally and ethically permitted.

In sum it can be held that the current law applies to the dehumanization of warfare and in general can govern the now foreseeable legal challenges sufficiently, if adapted accordingly with view to the aims and purposes of the rules. Thus, there is no need for new rules regulating the dehumanization of warfare. Nevertheless, there are (partly) high thresholds that have to be met by the technology. In case a weapon or weapon system is unable to comply with certain rules it cannot be introduced and used in a military operation. This is, however, depending on the particular tool and not on the existence of human controllability. It must be underlined that human control is not necessarily needed as a legal requirement for lawful behaviour, as it has been shown in the previous contributions.

In case a system cannot comply with some targeting rules but is not unlawful per se, still some missions without the risk of violation of LOAC could be lawful. This might be possible as the operations and their mission parameters itself could be adapted to the limited abilities of an autonomous system.

If a system has sophisticated autonomy its abilities could enable a better and lawful behaviour, which results possibly in less collateral damage and civil suffering. On the other hand higher developed autonomy could lead to unpredictable actions by the system. If the system control does not have fixed and insurmountable limits for compliance with the law or can override pre-determined parameters, there will be severe challenges for a use without violation of LOAC.

Especially the rules, which require value-based decision making, can hardly be inserted and simulated in a control system due to missing human judgment abilities. Already the necessary evaluation and value determination is—from the current technology-perspective–technically not possible. Therefore, the proportionality rules probably would be violated by an autonomous system. However, future artificial intelligence might be able to develop a value-based decision making similar to a human. This has to be identified as the decisive question, which lies at the bottom of the whole discussion about the legality of autonomous weapon system and cyber tools:

Is it possible to create such sophisticated autonomy or artificial intelligence, that it is able to comply with all relevant rules of law of armed conflict and the applicable international law?

Following the supporting opinion, the dismissal of compliance with the proportionality rule seems only to be temporarily and the decision could change in the future. Nevertheless for the moment the involvement of the human in form of an ability to override decisions by a (fully) autonomous system is a decisive factor for a lawful use of unmanned system and cyber tools.

However, critique from an ethical and moral standpoint remains. In case these caveats are considered as too big and irreconcilable by society, pressure has to be exerted on politics to create new rules and new treaties respecting these considerations. Other ways to enable an influence by moral and ethics in the law, e.g. by opening a door in the wording of a rule for serious ethical concerns cannot be identified and confirmed. Even if there might be general consent in different States about a change, this call also should not be followed with view to law of armed conflict. Such open rules of ethics would have to be abstract to be applicable to multiple situations, but their interpretation then would be open, too. Consequently it could happen that ethical concerns can be misinterpreted to the disadvantages of protected objects and persons resulting in an increase of civilian casualties and damages. Besides ethical perspectives can change over years, but a treaty cannot be changed that easily. Thereby "old" rules could be preserved in the law even if society and their concerns has changed to less or different requirements or other ethical problems.

If you would open this pandora's box the clearness and the functioning of the law is harmed as unforeseeable desires could be tried to be included in the law creating higher thresholds. Thereby, the will of States to comply with the legal framework will decrease. Also the authority for changing the law lies with the States and not singular groups with the loudest voices and their sometimes opaque interests.

Next, future development might show an ethical motivated counter-movement, which demands the use of fully autonomous systems, because their acts are more humane due to superior abilities and impossible rule transgression in comparison to the conduct of hostilities by human soldiers.

Drawing the line back to the introduction the existing law is applicable and already flexible enough to govern also the new (weapon) technologies effectively, which form the dehumanization of warfare. Nevertheless the discussion about missing human control in warfare, the conduct of hostilities and especially the targeting process is just beginning to evolve and couple of questions might pop up in the next few years.

Generally speaking, every technological development, especially in warfare, has always two faces. One face is formed by the negative consequences of faster and more efficient weapon systems with possible more grave effects. The suffering of civilians might aggravate and the situational awareness tends to become more blurred with more fog of war due to ever quicker and more distanced and possibly encoded processes and a massive amount of information. On the other side future systems and the communication have a bigger range, payload, endurance. Due to technological advances a better object recognition, quicker calculation processes and a more effective data analysis becomes possible. Increasing autonomy and AI leads to a more objective and eventually an independent decision making process through the weapon system. All of this leads to more efficiency and precision. Weapon systems entailing such abilities could theoretically reduce human suffering in warfare and decrease collateral damage. If a reduction of casualties due to high-performance technology is happening predictable and reliable consequently a duty of conflict parties to use such tools might result and become immanent. In such a scenario a bigger threat is posed by the human and not the machines. But still ethical questions remain at its core. For example, autonomous weapon systems cannot have feelings. So what to do about mercy in obvious situations like an attacking combatant? Of course, a lot of these situations could be resolved by equipping autonomous weapon systems or programming attack software only with tools which can create non-lethal effects. However, also non-lethal autonomous weapon systems or cyber tools could be criticized, because they might lower the (psychological) threshold of warfare as there are no "real" victims or they might foster the use of a weapon which is said to be unable to kill increasing violence against humans. Again: Every technology for warfare has its two faces and its original purpose might be perverted.

What can be concluded as the most important/decisive (core) element of the development of new weapon technologies is that the human always should have the final control about which weapon shall be introduced and used in warfare before it is used. This is incorporated in the rule of Art. 36 AP I. The central criterion for usability is the lawfulness of the weapon or weapon system. The legal review process is the last and decisive step to prevent particular dangerous or unforeseeable weapons that have to be viewed as unlawful. Therefore the importance of Art. 36 AP I must be stressed above all. Thus, all States should reiterate their commitment to Art. 36 AP I or the parallel customary international law and consequently should introduce an obligatory weapons review process on national level. Otherwise, it can be feared that premature and dangerous weapon technologies will be proliferated and used on the future battlefields with possible hazardous consequences for every human life on the ground.

On the other hand if high performance technology could be introduced in an organized and controlled way and is deployed according to the laws of war there will be less casualties and thereby a push for more humanity in warfare.

In contrast to conventional up-to-date warfare the progress will be felt significantly—especially by humans.

Index

W. Heintschel von Heinegg et al. (eds.), *Dehumanization of Warfare*,
DOI 10.1007/978-3-319-67266-3

CPSIA information can be obtained
at www.ICGtesting.com
Printed in the USA
LVHW01*2116080218
565816LV00002B/55/P

9 783319 672649